THERMAL CONDUCTION IN SEMICONDUCTORS

Thermal Conduction in Semiconductors

C.M. BHANDARI
Department of Physics
Allahabad University
Allahabad, India

D.M. ROWE
Department of Physics
Electronics and Electrical Engineering
University of Wales
Institute of Science and Technology
Cardiff, U.K.

WILEY EASTERN LIMITED
New Delhi Bangalore Bombay Calcutta Madras Hyderabad

Copyright © 1988, Wiley Eastern Limited
WILEY EASTERN LIMITED
4835/24 Ansari Road, Daryaganj, New Delhi 110 002
6 Shri B.P. Wadia Road, Basavangudi, Bangalore 560 004
Abid House, Dr. Bhadkamkar Marg, Bombay 400 007
40/8 Ballygunge Circular Road, Calcutta 700 019
Post Box No. 8604, Thiruvanmiyur, Madras 600 041
Post Box No. 1050, Himayath Nagar P.O., Hyderabad 500 029

This book or any part thereof may not be
reproduced in any form without the
written permission of the publisher

This book is not to be sold outside the
country to which it is consigned by
Wiley Eastern Limited

Rs. 150.00

ISBN 81-224-0064-7

Published by Mohinder Singh Sejwal for Wiley Eastern Limited,
4835/24 Ansari Road, Daryaganj, New Delhi 110 002 and printed
by Abhay Rastogi at Prabhat Press, 20/1 Nauchandi Grounds,
Meerut 250002.   Printed in India.

To

Rita and Barbara

It moves and it does not,
It is far yet it is near.
It is within all this, yet
It is also outside.

(Isha—Upanishad)

Preface

Over the past four decades or so material preparation techniques have been developed which have resulted in semiconductors becoming available with well-defined thermal and electrical properties. These new materials have facilitated the testing of the existing solid state theory and resulted in a better understanding of properties such as thermal conductivity. Studies of the variation of thermal conductivity with temperature and impurity concentration have provided a deeper insight into the nature of vibrational modes and various scattering processes that serve to limit the mean-free-path of heat carriers—phonons and electrons (holes). The availability of reliable thermal conductivity data has proved to be of considerable use in the selection of semiconductors for a wide range of applications such as thermoelectric generators/refrigerators, heat storage systems and photo-detectors.

A study of the thermal conductivity of semiconductors embraces a wide variety of topics and concepts. This book has been written to provide postgraduate students and researchers with a comprehensive coverage of the developments in this field, conveniently presented under one cover. An introduction and general survey of thermal conduction in solids presented in Chapter I is followed in Chapter II by an overview of the variety of methods employed in the measurement of thermal conductivity/ diffusivity of semiconductors. The principal contribution to the thermal conductivity of semiconductors, namely the electron (hole) and lattice contributions are discussed in the following four chapters. Electrons and holes are dealt with in Chapter III and electronic thermal conduction in Chapter IV; Chapter V covers lattice vibrations which is followed by Chapter VI on lattice thermal conductivity. Analysis of thermal conductivity data on elemental semiconductors is presented in Chapter VII and that on compound semiconductors in Chapter VIII. Amorphous and liquid semiconductors are dealt with in Chapter IX. Miscellaneous topics (magnetic and organic semiconductors and the effect of random inhomogeneities) are described in Chapter X. Finally the applications of thermal conductivity studies and other topics are covered in Chapter XI.

We are grateful to Professors H.J. Goldsmid, G.S. Verma, Bipin

Kumar Agarwal and J.E. Parrott for their help during the preparation of the book and Drs. Glen A. Slack, J.E. Aubrey, R.A.H. Hamilton and V.K. Agrawal for useful comments. Finally we thank our publishing manager Mr H.S. Poplai for his guidance during the preparation of the manuscript.

C.M. Bhandari
D.M. Rowe

January 1988

Acknowledgements

Chapters II, VII and VIII of this book are based upon a review paper "Preparation and Thermal Conductivity of Doped Semiconductors" (Prog. Crystal Growth and Charact., 1986, Vol. 13, pp. 233–289). For permission to reproduce in whole or in part, text, certain figures and diagrams we are grateful to Pergamon Journals Ltd. We are also grateful to the following publishers for permission to reproduce certain figures and tables.

John Wiley & Sons, Inc., Pergamon Press, American Institute of Physics, American Physical Society, The Institute of Physics, Plenum Publishing Corporation, Taylor & Francis Ltd., Academic Press, Inc., The Institute of Electrical & Electronic Engineers, Inc., Akademie—Verlag.

Detailed acknowledgements are given in the legends to figures.

List of Common Symbols

A	cross-sectional area, alloy disorder parameter in fine-grained material
a	interatomic spacing
$\mathbf{a_1}, \mathbf{a_2}, \mathbf{a_3}$	unit-cell vectors
$\mathbf{B}$	magnetic field
b	the magnitude of Burgers vector
C	specific heat
C_p, C_v	specific heats at constant pressure and constant volume
c	velocity of light
D	density-of-states function for electrons
D_a, D_b	density-of-states in acceptor and donor levels
E	energy, charge-carrier energy
E_g	energy band-gap
E_F	Fermi energy
$\Delta E_a, \Delta E_b$	energies corresponding to acceptor and donor levels with respect to the conduction band-edge
$\vec{\mathcal{E}}$	electric field intensity
e	electronic charge, subscript (electronic)
$\mathbf{F}$	force
F	Fermi integral
$f_\mathbf{k}$	distribution function for electrons
f	Fermi function
$\mathbf{G}$	reciprocal lattice vector
$g(v)$	density-of-states function for phonons

g	spectroscopic splitting factor
H	Hamiltonian operator
$\mathbf{h}$	heat flow vector
h	Planck's constant
$\hbar$	$h/2\pi$
j	electric current density
$\mathbf{k}$	electron (or hole) wavevector
k	magnitude of $\mathbf{k}$
k_B	Boltzmann constant
L	Lorenz number
$\mathcal{L}$	Lorenz factor
L_0	Lorenz number in the degenerate limit
${}^nL_l^m$	generalized Fermi integrals
l	mean-free-path
M	atomic mass
$\overline{M}$	mean atomic mass
m_0	free electron mass
m_d^*	density-of-states effective-mass
m_c^*	conductivity or inertial effective-mass
$m_\parallel^*, m_\perp^*$	effective-mass components parallel and perpendicular to the principal axis
N_q	Phonon distribution function
n	carrier concentration, refractive index
P_E	Ettinghausen coefficient
Q_N	Nernst coefficient
$\mathbf{q}$	phonon wavevector
q	magnitude of $\mathbf{q}$
R	gas constant
$\mathbf{R}_n$	lattice translation vector
r	radial coordinate
S	entropy
S_{RL}	Righi-Leduc coefficient
s	polarisation index, scattering parameter

LIST OF COMMON SYMBOLS

T	absolute temperature
T	tesla
t	time
U	crystal potential energy
u	atomic displacement
V	volume
v, v_s	sound or phonon velocity
W	thermal resistivity
x	position coordinate, $\hbar\omega/k_B T$
α	Seebeck coefficient, absorption coefficient
β	$k_B T/E_g$ (inverse of the reduced energy band-gap), force constant
β'	material parameter
Γ	disorder parameter
γ	Gruneisen constant
γ_∞	Gruneisen constant at high temperature
$\tilde{\gamma}$	γ appropriate to acoustic branch
Δ	related to chemical-shift (4Δ)
δ	Dirac delta function
$\epsilon_0, \epsilon_\infty$	static and high-frequency dielectric constants
ξ	$E_F/k_B T$, reduced Fermi energy
η	$E/k_B T$, reduced carrier energy
θ	temperature
θ_D	Debye temperature
$\theta_{DO}, \theta_{D\infty}, \tilde{\theta}_{D\infty}$	Debye temperature at $0\,K$, its high temperature limit and $\theta_{D\infty}$ value appropriate to acoustic phonons
χ	compressibility
λ	thermal conductivity
λ_e, λ_b	electronic (polar) thermal conductivity, bipolar thermal conductivity
$\lambda_L, \lambda_m, \lambda_{photon}$	lattice, magnon and photon contributions to λ
λ'	thermal diffusivity
μ	mobility
μ_B	Bohr magneton

xvi LIST OF COMMON SYMBOLS

ν	frequency
ρ	electrical resistivity, density
σ	electrical conductivity
σ_0	Stefan-Boltzmann constant
τ	relaxation time
ϕ	volume fraction
ω	angular frequency
ω_D	Debye frequency

Contents

Preface .... *ix*
Acknowledgements .... *xi*
List of Common Symbols .... *xiii*

1. **Introduction and General Survey** 1
   1.1 Introduction *1*
   1.2 Thermal conductivity values: metals, insulators and semiconductors *2*
   1.3 Definition of thermal conductivity—one-dimensional flow of heat *7*
   1.4 Insulators *8*
   1.5 Semiconductors *9*

2. **Measurement of Thermal Conductivity** 12
   2.1 General considerations *12*
   2.2 Static (steady-state) methods *15*
   2.3 Non-steady-state (dynamic) methods *20*
   2.4 Miscellaneous methods *27*

3. **Electrons and Holes in Semiconductors** 37
   3.1 Introduction *37*
   3.2 Electrons in solids *37*
   3.3 Metals, insulators and semiconductors *40*
   3.4 Electrons and holes *41*
   3.5 Energy bands in real crystals *43*
   3.6 The Fermi level *44*
   3.7 Electrons as carriers of charge and heat *47*
   3.8 Transport processes in solids *47*
   3.9 Electron Boltzmann equation and its solution *49*
   3.10 Electron scattering mechanisms *52*

4. **Electronic Thermal Conductivity** 59
   4.1 Introduction *59*

4.2 Extrinsic semiconductors—evaluation of various electronic transport coefficients using the relaxation time approximation 59
4.3 Evaluation of the Lorenz factor $\mathcal{L}$ 61
4.4 Mixed scattering: acoustic scattering and ionized-impurity scattering operative simultaneously 62
4.5 Intrinsic semiconduction 63
4.6 Nonparabolic nature of energy bands 66
4.7 Thermomagnetic effects 71

5. Phonons and Phonon Scattering Processes 78
5.1 Introduction 78
5.2 Vibrations in a one-dimensional crystal 78
5.3 Real crystals 79
5.4 Effect of defects—lattice imperfections 82
5.5 Vibrational properties of non-crystalline solids 83
5.6 Phonon Boltzmann equation—relaxation time 83
5.7 Scattering of phonons 84
5.8 Relaxation times for three-phonon processes 87
5.9 Other phonon scattering mechanisms—scattering by defects in the lattice 89
5.10 Scattering of phonons by electrons (or holes) 92

6. Lattice Thermal Conductivity 99
6.1 Introduction 99
6.2 The variational method 99
6.3 Absolute magnitudes 105
6.4 Lattice thermal conductivity—the relaxation time method 107
6.5 Normal processes 109
6.6 Generalization of Klemens-Callaway expression 110
6.7 Variation of λ with temperature 111
6.8 The $1/T$ law 114
6.9 Doped semiconductors 115
6.10 Imperfect crystals 116
6.11 Minimum of thermal conductivity 117
6.12 Method of Guyer and Krumhansl 120
6.13 Other carriers of heat 121

7. Semiconducting Materials—Analysis of Experimental Data (I) 131
7.1 Introduction 131
7.2 Group IV elements—germanium and silicon 132
7.3 Alloys of silicon and germanium 137
7.4 Radiative heat transfer 139
7.5 Fine-grained silicon-germanium alloys 142

7.6 Other elemental semiconductors *143*
7.7 Concluding remarks *146*

8. **Semiconducting Materials—Analysis of Experimental Data (II)** 149
   8.1 Introduction *149*
   8.2 Thermal conductivity of III-V compounds *149*
   8.3 III-V mixed crystals *158*
   8.4 Lead chalcogenides *160*
   8.5 II-VI compounds *164*
   8.6 II-IV compounds *166*
   8.7 Magnetic field effects *166*
   8.8 Bismuth telluride and other V-VI compounds *167*
   8.9 Other semiconductors *171*

9. **Amorphous and Liquid Semiconductors** 175
   9.1 Amorphous materials *175*
   9.2 Liquid semiconductors *183*

10. **Miscellaneous Semiconductors** 192
    10.1 Introduction *192*
    10.2 Magnetic semiconductors *192*
    10.3 Organic semiconductors *198*
    10.4 Heterogeneous solids *200*

11. **Applications of Thermal Conductivity Studies and Other Topics** 209
    11.1 Introduction *209*
    11.2 Practical importance of thermal conductivity studies in semiconductors *209*
    11.3 Dopant precipitation in heavily doped semiconductors *214*
    11.4 Nonmetallic crystals with high thermal conductivity *214*
    11.5 Thermal conductivity at high pressure *215*
    11.6 Radiation damage in solids—effect on thermal conductivity *217*

    Subject Index 223

7.7. Other elemental semiconductors 144
7.7. Concluding remarks 146

8. Semiconducting Materials—Analysis of Experimental Data (II) 149
   8.1. Introduction 149
   8.2. Thermal conductivity of III–V compounds 149
   8.3. III–V mixed crystals 158
   8.4. Solid solid-solutions 160
   8.5. II–VI compounds 164
   8.6. II–IV compounds 166
   8.7. Mixed materials 168
   8.8. Organic solids and other V–VI compounds 162
   8.9. Thermal conductivity 171

9. Amorphous and Liquid Semiconductors 173
   9.1. Amorphous Substances 173
   9.2. Liquid semiconductors 174

10. Miscellaneous Substances 177
    10.1. Introduction 177
    10.2. Molecular semiconductors 178
    10.3. Organic compounds 179
    10.4. Heterogeneous solids 200

11. Applications of Thermal Conductivity Studies and Other Topics 209
    11.1. Introduction 209
    11.2. Practical importance of thermal conductivity studies on semiconductors 209
    11.3. Dopant precipitation in heavily doped semiconductors 214
    11.4. Nonmetallic crystals with high thermal conductivity 214
    11.5. Thermal conductivity at high pressure 215
    11.6. Radiation damage in solids—effect on thermal conductivity 217

Subject Index 223

Chapter 1

Introduction and General Survey

1.1 Introduction

It is only during recent years that heat transport mechanisms in solids have been intensively studied. The development of semiconductor technology in the 1950's was accompanied by a requirement for information on a variety of fundamental material parameters. "Perfect" and "pure" semiconductor crystals became available and these substances provided an excellent opportunity to investigate in detail a number of physical processes in solids.

At moderate temperatures, the thermal transport behaviour of a pure (intrinsic) semiconductor is similar to that of an insulator with heat conduction due to lattice waves (phonons). Controlled amounts of suitable impurities (dopants) can be added to a semiconductor producing electrons or holes in desired numbers. These charge carriers give rise to an "electronic" contribution to thermal conductivity. However, apart from serving as heat carriers, electrons and holes may also act as scattering centres for phonons and cause a reduction in lattice thermal conductivity. At temperatures sufficiently high to excite carriers across the semiconductor energy band gap, electron-hole pairs transport heat and give rise to a bipolar contribution to the thermal conductivity.

Thus the total thermal conductivity λ of a semiconductor can be generally represented by

$$\lambda = \lambda_L + \lambda_e + \lambda_b \qquad (1.1)$$

where λ_L, λ_e and λ_b refer to the lattice, electronic (polar) and bipolar contributions, respectively.

Other contributions to thermal conductivity may become significant in specific situations: for example, the conduction of heat by magnons (quanta of spin waves) in ferromagnetic materials at very low temperatures, while at high temperatures radiative heat transfer (photon thermal conduction) may be appreciable in materials such as tellurium, selenium and silicon-germanium alloys. There have also been suggestions of heat transport by excitons although, to date, no definite proof of an appreciable excitonic contribution has been obtained.

1.2 Thermal conductivity values: metals, insulators and semiconductors

Semiconductors possess electrical resistivities which are intermediate between those of metals and insulators (generally in the range $10^{-4}-10^7$ Ωm at room temperature). Evidently, it is relevant to examine the range over which thermal conductivity varies from metals to insulators and semiconductors. The thermal conductivity of a solid is sensitive to the nature and concentration of impurities and imperfections. Consequently, comparisons are drawn between the thermal conductivity values of undoped single crystal materials.

Pure copper has a thermal conductivity of about 400 Wm^{-1}K^{-1} at room temperature. Only a small fraction of this thermal conductivity (about 1 or 2 per cent) arises from the lattice contribution. λ_L (calc) = 5 Wm^{-1}K^{-1}; the high concentration of electrons (as compared with non-metallic solids) being responsible for the large electronic contribution. Pure, undoped silicon, on the other hand, has a thermal conductivity of 145 Wm^{-1}K^{-1}, all of which is the lattice contribution. The small lattice contribution in metals can be explained on the basis of a significant scattering of lattice waves (phonons) by electrons.

Although metals are in general better conductors of heat than non-metals, some non-metallic substances possess thermal conductivities comparable to those of metals. Berman et al. (1951) have shown that, at room temperature, diamond conducts heat better than copper or silver. Amongst metals, silver has the highest thermal conductivity (430 Wm^{-1}K^{-1}) at room temperature. Non-metallic substances with thermal conductivities exceeding 100 Wm^{-1}K^{-1} may be termed as high thermal-conductivity materials. However, only a few nonmetallic crystals fall into this category with most of them, such as diamond, silicon carbide, silicon, beryllium oxide, gallium phosphide, etc. having adamantine structures. Other materials in this category are: AlN, AlP, BAs, cubic BN, BP, BeS and GaN, the conditions which give rise to such high thermal conductivities in non-metallic crystals have been discussed by Slack (1973) and will be described in Chap. XI.

Table 1.1 compares the room-temperature thermal conductivities of various semiconductors. Some metals and insulators have also been included for comparison.

Apart from the low-conductivity composite materials, the semiconductors with lowest values of thermal conductivity are Cd_3As_2, Ga_2Se_3, $AgSbS_2$ and $AgSbTe_2$ having thermal conductivities in the range 0.3-0.6 Wm^{-1}k^{-1}. Copper, silver and diamond have thermal conductivities more than three orders of magnitude higher. The electrical resistivities of various solids show a much greater range ($10^{-8}\Omega$m for good conductors to 10^{20} Ωm for insulators). The fact that an absence of free charge carriers in insulators and

Table 1.1
Thermal Conductivity of Various Solids at Room Temperature

Group	Material	Thermal Conductivity* ($Wm^{-1} K^{-1}$)	Reference Source
(1)	(2)	(3)	
Metals	copper	400	
	silver	430	
Semimetals	bismuth	8.5 }	(Kaye and Laby, 1966)
	antimony	18 }	
Semiconductors			
Group IV	diamond	2000	(Berman et al, 1956)
	silicon	145 }	(Carruthers et al, 1957)
	germanium	64 }	
Group VI	selenium	2	(White et al, 1958)
	tellurium	3	(Fischer et al, 1957)
III–V	GaP	110 }	(Weiss, 1959)
	InP	80 }	
	AlSb	60 }	
	GaAs	37	(Abrahams et al, 1959)
	InAs	29	
	GaSb	27	(Wright, 1959)
	InSb	15	(Stuckes, 1957)
II–IV	Mg_2Si	10.5 }	(Martin, 1972)
	Mg_2Ge	13.0 }	
	Mg_2Sn	16.0 }	
II–V	ZnSb	1.0 }	(Turner et al, 1961)
	CdSb	1.1 }	
	Cd_3As_2	0.3–0.4	(Spitzer et al, 1966)
II–VI	BeO	370	(Slack and Austerman 1971)
	HgTe	2	(Wright, 1959)
	CdSe	9	(Jacobs and Irey,[1] 1970)
	CdTe	7.5	(Slack and Galginaitis, 1964)

(Contd.)

4 THERMAL CONDUCTION IN SEMICONDUCTORS

(1)	(2)	(3)	
	CdS	20	(Holland, 1964)
	ZnTe	18	(Slack, 1972)
III–VI	Ga_2Se_3	0.5	(Ioffe, 1959)
IV–VI	PbSe	1.7	(Ioffe, 1956)
	PbS	3.0	(Ioffe, 1960)
	PbTe	2.3	(Ioffe, 1956)
	GeTe	1–3	(Miller, 1961)
V–VI	Sb_2Te_3	2.4 ⎫	(Birkholz, 1958)
	Bi_2Se_3	2.4 ⎭	
	Bi_2Te_3	1.6	(Goldsmid, 1958)
I–III–VI	$AgInSe_2$	3.0	(Wright, 1959)
	$CuInSe_2$	3.7	(Davisson and Posternak, 1962)
I–V–VI	$AgSbTe_2$	0.6	(Hockings, 1959)
	$AgSbS_2$	0.5	(Rosi et al, 1960)

Other non-metallic solids with high thermal conductivities

	cubic BN	1300[2]	
	BAs	210	(Slack, 1973)
	GaN	170	
	AlP	130	

Other materials[3]

Nichrome ($Ni_{.70}Cr_{.30}$)	14	
Glass	0.8	
Polystyrene	0.1	
Expanded polystyrene	0.035	(density 16 kg m^{-3})
Rubber	0.2	
Concrete	0.9	
Dry wood	0.17	(density 800 kg m^{-3})

1. Estimated, extrapolated from 100 K
2. Estimated value for pure large crystals
3. See for example, Pratt (1969) and Parrott and Stuckes (1975)

*Units: The S.I. system of units have been generally used in the book. For thermal conductivity this unit is W m^{-1} K^{-1} (or W m^{-1} deg^{-1}).

undoped semiconductors at low temperatures is responsible for their low electrical conductivity is readily understood The lattice thermal conductivity in these materials is the only important contribution to the thermal conductivity and this depends upon the atomic mass, nature of binding and the degree of perfection of the specimen under study which in turn depend upon the absence of any impurities, disorder or boundaries.

In general the role of electrons (or holes) as carriers of heat and also as scattering centres for phonons appears to explain, in a qualitative way, the range over which the thermal conductivity values are spread in various classes of materials. Figure 1.1 gives an estimate of the order of magnitude of carrier concentrations for metals, semimetals and semiconductors.

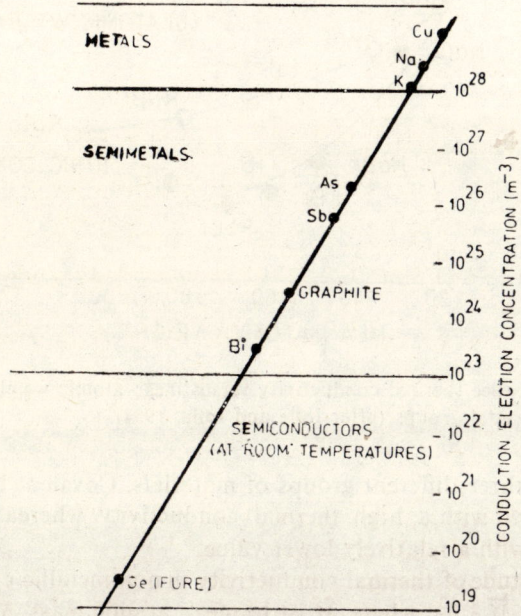

Fig. 1.1 Concentration of carriers in different solids. The semiconductor range may be extended upwards by suitable doping (after Kittel 1976; reprinted by permission of John Wiley & Sons. Ltd).

The theory of thermal conductivity has received considerable attention during the last three decades and it is now possible to provide a theoretical explanation of the thermal conductivity behaviour of solids over a considerably wide range of temperatures. The two theoretical approaches—the relaxation-time approach and the variational approach—have their own strengths and weaknesses and are to be discussed in Chaps. III and VI.

The variation of thermal conductivity of non-metallic solids with atomic weight and upon the type of binding shows that, for a group of similar crystals, the lattice thermal conductivity falls with increasing atomic weight (or mean atomic weight for compounds). Figure 1.2 shows this

6 THERMAL CONDUCTION IN SEMICONDUCTORS

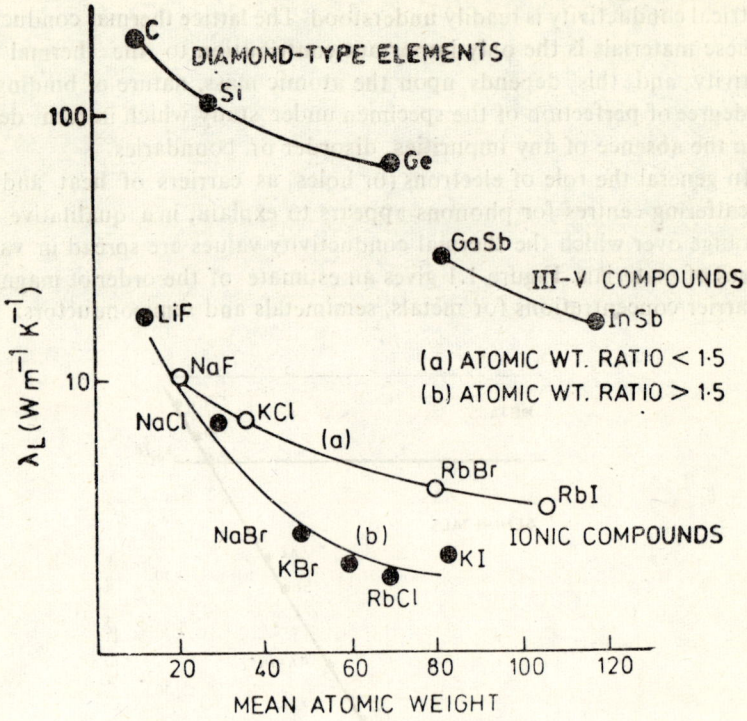

Fig. 1.2 Lattice thermal conductivity versus mean atomic weight for various crystal groups (after Ioffe and Ioffe, 1954).

variation for three different groups of materials. Covalent binding appears to be associated with a high thermal conductivity whereas ionic binding is associated with a relatively lower value.

The magnitude of thermal conductivity in non-metallic crystals depends upon a factor $\overline{M} \delta \theta_D^3$, where $\overline{M}$ ist he mean atomic mass, δ^3 is the average volume occupied by one atom and θ_D is the Debye temperature. The factor $\overline{M} \delta \theta_D^3$ is maximized for light mass and strongly bonded crystals (see Chap. XI).

Liquid and non-crystalline materials also exhibit semiconducting properties. Some organic materials also show semiconducting properties although their thermal conductivity behaviour has so far not been discussed in great detail. The thermal properties of amorphous solids at low temperatures (below about 50 K) show a number of unusual features not observed in corresponding crystalline counterparts. There is very little difference between the thermal conductivities of the various amorphous solids compared to those between different crystals. On these grounds there have been suggestions for the use of some amorphous materials as thermal conductivity standards (see Chap. IX).

1.3 Definition of thermal conductivity—one-dimensional flow of heat

The definition of heat conduction assumes that the rate of transfer of heat by conduction in a solid across an element of surface is proportional to the area of the surface element, the temperature gradient ($d\theta/dx$) at that point and the cosine of the angle between the normal (to the surface element) and the temperature gradient. The coefficient of the thermal conductivity λ is then defined by the following equation:

$$\dot{Q} = -\lambda A \left(\frac{d\theta}{dx}\right) \cos \phi \tag{1.2}$$

Here $\theta(x)$ refers to the temperature measured in arbitrary units and $\dot{Q}$ is the rate of heat transfer across the element of area A whose normal makes an angle ϕ with the temperature gradient. Usually A is chosen normal to the temperature gradient; the cosine term is then unity and $\dot{Q}$ is expressed as $-\lambda A (d\theta/dx)$.

Most of the early investigations of the thermal conductivity of solids started with the study of metals. Based on the principle of energy conservation, the amount of heat flowing across a certain cross-section can be written as a sum of the heat absorbed by a section of the material and the heat lost by radiation. Material is usually considered in the form of a cylindrical bar with heat flowing across its length which may be assumed to align with the x-direction. If $\theta(x)$ refers to the temperature (at any point x along the bar) in excess of that of the surrounding medium, one obtains

$$\frac{d\theta}{dt} = \lambda' \left(\frac{d^2\theta}{dx^2}\right) - \mu'\theta \tag{1.3}$$

where $\lambda' = \lambda/C$, is the thermal diffusivity of the material and the parameter μ' depends upon the perimeter of the cross-section of the bar, emissivity of its surface, cross-sectional area A and the specific heat per unit volume C (see for example, Saha and Srivastava, 1934, Carslaw and Jaeger, 1959).

This is the basic equation which gives the temperature distribution along the length of the bar. In the steady state, $d\theta/dt = 0$ at all points. With a given set of boundary conditions, the equation can be solved and temperature obtained as a function of x. Fourier described a method of solving the equation and his method has been widely adopted.

The definition of thermal conductivity as given above must be supplemented with the condition of zero current flow. The charge carriers (electrons and holes) may contribute significantly to the total heat flux in semiconductors. In this description of one-dimensional flow of heat, the thermal current density w_x can be expressed in terms of the electric current density j_x and the temperature gradient $\partial\theta/\partial x$. Obviously, λ can then be related to the thermal current density per unit temperature gradient if $j_x = 0$. Chapter III gives a relatively detailed account of the transport equations and various transport coefficients. The experimental methods for the measurement of thermal conductivity are described in Chapter II.

In metals almost all of the heat is carried by electrons. The lattice contribution, although generally small, can be appreciable in specific situations. In semi-metals, such as bismuth and antimony, where the carrier concentration is low in comparison with metals, the lattice component of thermal conductivity becomes appreciable.

1.4 Insulators

In 1911 Eucken analysed the experimental data of a number of insulators and concluded that $\lambda \propto T^{-1}$ (where T is the temperature) for dielectric crystals. This observation refers to the high temperature behaviour (high as compared with the Debye temperature θ_D). Clearly, any theoretical model should be able to explain this T^{-1} dependence of thermal conductivity.

In a solid atoms are held in their positions in the lattice by interatomic forces. The atoms do not, in general, possess translational or rotational degrees of freedom but can vibrate about their mean positions. It is possible to relate the temperature of the solid to the amplitude of the atomic vibration, higher temperature corresponding to a higher amplitude. The more strongly vibrating atoms tend to pass some of their energy to their nearest neighbours which in turn will pass some of it to their neighbours. If the material is held (say in the form of a disc) with one face in contact with a heat source and the other in contact with a heat sink heat is then conducted from one face to the other.

A different approach towards an understanding of the mechanism of heat conduction would be to think in terms of the collective vibrations of the whole system rather than those of the individual atoms. In a solid any displacement given to a particular atom does not remain localized, instead it travels through the lattice giving rise to what is known as a lattice wave or a displacement wave. The thermal conduction can then be described as the propagation of energy through the crystal by these lattice waves. These lattice waves can be quantized and the resulting quantum is referred to as a phonon. This subject will be discussed in Chap. V.

Many thermal properties of solids can be described in terms of phonon behaviour. In a perfect crystal (absence of any imperfections, impurities or boundaries) and in the harmonic approximation (retaining terms only up to second order in the expansion of the crystal potential) the phonon modes propagate without any interference and are referred to as Normal modes. However, a certain amount of anharmonicity (terms beyond the second order in the expansion of the crystal potential) is always present and this leads to the scattering of a phonon by other phonons. In imperfect crystals, phonons may also be scattered by other mechanisms, such as impurities, lattice imperfections, boundaries, etc. In heavily doped semiconductors the scattering of phonons by electrons and holes may also contribute to the total phonon scattering. Associating a mean-free-path l with the phonons the expression for the lattice thermal conductivity can be written as

$$\lambda_L = \frac{1}{3} C \, v_s \, l \qquad (1.4)$$

C is the specific heat per unit volume and v_s the sound (phonon) velocity. As the scattering of phonons increases the phonon mean-free-path l decreases. The variation of λ_L with temperature will then mainly depend upon the temperature variation of the product Cl as v_s changes only slightly with temperature.

Peierls (1929) obtained a T^{-1} dependence of the thermal conductivity above the Debye temperature on the basis of his theory and explained Eucken's experimental T^{-1} law. As the temperature is lowered λ_L rises and reaches a maximum at a temperature well below the Debye temperature. With a further lowering of temperature λ_L decreases rapidly and vanishes at absolute zero. A simple explanation of this behaviour was given by Casimir (1938). As the temperature decreases the phonon mean-free-path increases rapidly, resulting in an increase in λ_L until it becomes comparable to the crystal dimensions. The boundaries of crystals are usually poor reflectors of phonons and, consequently, their mean-free-paths cannot increase any further. In this temperature range the specific heat (of the lattice) follows a T^3 law. The lattice thermal conductivity now decreases with a decrease in temperature following the same T^3 law.

1.5 Semiconductors

The ideas presented in Sec. 1.4 describe the state of affairs which exists in crystalline non-metallic solids. This includes a pure undoped semiconductor at low temperatures. The introduction of impurities into the lattice brings about changes in the phonon spectrum and the impurities may also act as scattering centres for phonons. The doping of a semiconductor introduces electrons and holes in the conduction and valence bands which can act as carriers of heat. On the other hand, they may also act as scattering centres for phonons, and the phonon-electron interaction plays an important part in limiting the mean-free-paths of both types of heat carriers.

The introduction of certain impurities in the host lattice can give rise to new vibrational modes in the gap region and also in the allowed energy region of the host crystal. These modes have a significant effect on low-temperature thermal conductivity (see Chap. V).

The electronic polar (either electrons or holes) contribution to thermal conductivity is usually written as

$$\lambda_e = L \, \sigma \, T \tag{1.5}$$

L, which is similar to the Lorenz number for metals, acquires a value of around $2 \, (k_B/e)^2$ in the non-degenerate region (if the electrons are assumed to be scattered by acoustic phonons only), whereas in the degenerate limit L acquires a value $L_0 = \dfrac{\pi^2}{3} (k_B/e)^2$. A detailed discussion on the theoretical calculation of L is presented in Chap. IV.

In the intrinsic region, the bipolar thermal conductivity becomes significant when electron-hole pairs across the energy-band gap contribute to heat conduction as they flow down the temperature gradient and recombine to release the energy of ionization. The Lorenz number associated with this process may be much higher than its value when only one type of carrier (electron or hole) contributes to heat conduction.

In a number of small gap semiconductors the theoretically calculated values of L may be much larger than can be accounted for on the basis of experimentally determined values of electronic thermal conductivity. This discrepancy has been attributed to the fact that the energy bands are not essentially parabolic. The non-parabolicity of energy bands has a considerable influence on various electronic transport parameters. Ravich *et al* (1971) have discussed this aspect of the problem with reference to the transport properties of lead chalcogenides.

Semiconductors provide an excellent opportunity for testing the basic concepts of the various processes which contribute to heat conduction and the different mechanisms related to the scattering of phonons and electrons. On the other hand theoretical predictions can be utilized in developing materials with specific properties which are tailored for specialized applications. In a number of situations it is important to know the thermal conductivity of the material (see Chap. XI). In some devices, such as thermoelectric generators and refrigerators, a low lattice thermal conductivity is required. In other situations, where there is large internal power dissipation, a high thermal conductivity material is required to conduct away heat to the sink. These transport property requirements are elaborated in Chap. XI.

References

Abrahams, M.S., Braunstein, R. and Rosi, F.D. (1959) *J. Phys. Chem. Solids 10*, 204.

Berman., R.,Foster, E.L. and Ziman, J.M. (1956), *Proc. Roy. Soc.* London A 237, 344.

Berman, R., Simon, F.E. and Wilks, J. (1951), *Nature* 168, 277.

Birkholz, U. (1958) *Z. Naturforsch 13 a*, 780.

Carruthers, J.A., Geballe, T.H., Rosenberg, H.M. and Ziman, J.M. (1957), *Proc. Roy. Soc.* London A 238, 502.

Carslaw, H.S. and Jaeger, J.C. (1959), *Conduction of Heat in Solids*, Clarendon Press, Oxford, edition II.

Casimir, H.B G. (1938), *Physica 5*, 495.

Davisson, J.W. and Posternak, J. (1962), *Status Report on Thermoelectricity*, NRL Mem. Report 1241, p. 30.

Eucken, A. (1911), *Ann. Physik* 34, 185.

Fischer, G., White, G.K. and Woods, S.B. (1957), *Phys. Rev.* 106, 480.

Goldsmid, H.J. (1958), *Proc. Phys. Soc.*, London 72, 17.

Hockings, E.F. (1959), *J. Phys. Chem. Solids 10*, 341.

Holland, M.G. (1964), *Phys. Rev.* 134, A 471.

Ioffe, A.F. (1956), *Can.J. Phys.* 34, 1342.

Ioffe, A.F. (1959), Sov. *Phys.-Solid St. 1.* 141.

Ioffe, A.F. (1960), Physics of Semiconductors, *Infosearch*, London.

Ioffe, A.V. and Ioffe, A.F. (1954), *Dokl Akad. Nauk. SSR.* 97, 821.

Jacobs, R.I. and Irey, R.K. (1970), *Proc. 9th Conf. on Thermal Cond.* (ed. H.R. Shanks), U.S. AEC, Oak Ridge, Tenn., p. 25.

Kaye, G.W.C. and Laby, T.H. (1966), *Table of Physical and Chemical Constants*, Longman Green, London.

Kittle, C. (1976), *Introduction to Solid State Physics*, John Wiley, New York (5th Ed).

Martin, J.J. (1972), *J. Phys. Chem. Solids. 33*, 1139.

Miller, R.C. (1961), *Thermoelectricity: Science and Engineering* (ed. R.R. Heikes and R.W. Ure) Interscience, New York, p. 438.

Olsen, J.L. (1962), *Electron Transport in Metals*, Interscience, New York.

Parrott, J.E. and Stuckes, A.D. (1975), *Thermal Conductivity of Solids*, Pion Limited, London.

Peierls, R.E. (1929), Ann. Physik *3*, 1055.

Pratt, A.W. (1969), *Thermal Conductivity* (ed. R.P. Tye), Vol. 1, Academic Press.

Ravich, Yu, I., Efimova, B.A. and Tamarchenko, V.I. (1971), *Phys. State Solidi* (b) 43, 11.

Rosi, F.D., Dismukes, J. P. and Hockings, E. F. (1960), *Elect. Engng.* 79, 430.

Rowe, D.M. and Bhandari, C.M. (1983), *Modern Thermoelectrics*, Holt-Saunders, Lond.

Saha, M.N. and Srivastava, B.N. (1934), *A Treatise on Heat*, Indian Press Limited.

Slack, G.A. (1972), *Phys. Rev., B6*, 3791.

Slack, G.A. (1973), *J. Phys. Chem. Solids* 34, 321.

Slack, G.A. and Austerman, S.B. (1971), *J. Appl. Phys.* 42, 470.

Slack, G.A. and Galginaitis, S. (1964), *Phys. Rev.* 133, A 253.

Spitzer, H.P., Costellion, G.A. and Haacke, G. (1966), *J. Appl. Phys.* 37, 3792.

Stuckes, A.D. (1957), *Phys. Rev. 107*, 427.

Turner, W.J., Fischer, A.B. and Reese, W.E. (1963), *Phys. Rev.* 121, 750.

Weiss, H. (1959), *Ann. Physik 4*, 121.

White, G.K., Woods, S.B. and Elford, M.T. (1958), *Phys. Rev.* 112, 111.

Wright, D.A. (1959), *Electronic Engng.* 31, 659.

Chapter 2

Measurement of Thermal Conductivity

2.1 General considerations

2.1.1 *Introduction*

Accurate measurement of thermal conductivity λ is critically dependent upon preventing unwanted heat transfer between the surface of the sample under examination and the surrounding medium. Any interchange of heat between the two will distort the geometry of the isothermals which satisfy the boundary conditions appropriate to the method of measurement employed. In practice this proves difficult because, unlike in electrical conductivity measurements in which electric current (flow) losses along the length of the sample can be prevented by suitable electric insulation, such as air, it is very difficult to establish a similar loss-less situation in the measurement of thermal conductivity. However, heat exchange mechanisms are understood and heat transfer can be significantly reduced by making measurements in vacuum and through the use of radiation shields and thermal insulation. Similar heat transfer problems are encountered in the measurement of thermal diffusivity, although not to the same extent.

Thermal conductivity is sensitive to small changes or differences in the physical properties of the material such as grain size, homogeneity, porosity, etc. and widely differing values of thermal conductivity have been reported by different workers for the same material (see for example, *"Thermophysical properties of matter'* **Y.**S. Touloukian, ed.). Although a portion of the discrepancy can be attributed to differences in the physico/chemical properties of the samples investigated, the major cause of discrepancy is the different extent to which investigators satisfy experimentally the required heat flow conditions appropriate to their method of measurement. Thermal conductivity apparatuses are notorious for systematic errors and their performances should, if possible, be checked over the intended temperature range of measurement by carrying out thermal conductivity measurements on appropriate "standard" materials. Unfortunately this is usually not possible in the case of most semiconductors. Materials currently stocked in the US National Bureau of Standards as thermal conductivity standards are tungsten, stainless steel and electrolytic iron. There are materials, such as pyrex 7740 and pyrocerm 9606 which are useful when making measurements of λ in the range $1 < \lambda < 6$ W/mK. Although these materials have not yet been designated as standards they have been subjected to round robin testing and are being archived (Tye and Hulstom, 1984).

Descriptions of apparatus and discussions of the various methods employed in the measurement of thermal conductivity of solid and liquid semiconductors can be found in many texts (Drabble and Goldsmid, 1961, Tye, 1969, Regel et al., 1971, Parrott and Stukes, 1975 and Berman, 1976). This section provides an updated and collective overview of the state-of-the art.

2.1.2 Measuring techniques

The methods generally adopted in the measurement of the thermal conductivity of semiconductors do not differ in principle from the methods used for other solids or liquids. A notable exception is the Harman's method employed in the measurement on thermoelectric materials (Harman, 1958) in which the Peltier effect is used to establish a temperature gradient along the sample. All methods involve either the direct measurement of heat which passes through the sample (absolute methods) or comparing the amounts of heat flowing through the sample under examination with that flowing through a sample of known thermal conductivity (comparison methods).

Thermal conductivity measurement techniques can be divided into two groups, static methods and non-steady or dynamic methods. In static methods, measurements are made only after thermal equilibrium has been established. Although time consuming, such methods guarantee achievement of high accuracies. On the other hand, in dynamic methods, thermal gradients are observed as a function of time, enabling measurements to be made relatively quickly and over a wide range of temperatures. However, in dynamic methods it is usually the thermal diffusivity which is measured; the specific heat and density of the sample material must be known in order to obtain the thermal conductivity.

The choice of method is governed initially by the intended temperature range of measurement. Static methods are employed at low temperatures and dynamic methods at room temperature and above. Other factors which should be taken into account are; sample size, geometry, required speed of measurement, and accuracy.

2.1.3 Measurement of temperature

The measurement of thermal conductivity or thermal diffusivity of a material depends upon the use of thermometers to measure either temperature gradients or a change in temperature with time. Choice of thermometers depends upon a number of factors such as the method employed in the measurement of the thermal properties, the temperature range of measurement and the required accuracy of the measurement. A comprehensive account of the measurement of temperature has been given by Quinn (1983) and reviewed by Crovini (1984).

Originally, gas thermometers were used in low-temperature measurements on semiconductors (White, 1969) but they were superseded by carbon and semiconductor resistance thermometers (Barber, 1960) although carbon thermometers are unstable to thermal cycling and exhibit non-monotonic magnetoresistance at very low temperatures and in high fields (Naughton et al, 1983).

Thermocouples are probably the most widely used devices for measuring temperature (Billing et al, 1975, Benedict, 1977, and Broomfield, 1979). The Seebeck (thermoelectric) coefficients of the thermocouple materials, however, are sensitive to mechanical strain and contamination. Consequently, care should be taken to avoid the distortion of thermocouple wires more than is absolutely necessary and they should be annealed before use. Contamination of the thermocouple by the semiconductor material or vice versa can often be prevented by providing a suitable diffusion barrier. Thermocouples fabricated from alloys such as gold/2.1% cobalt versus copper (Lindenfield, 1962), manganin (Slack, 1961) or normal silver (Holland and Rubin, 1962) have been widely used in low-temperature measurements on semiconductors, but the gold/2.1% cobalt (negative) alloy was found to be unstable when thermally cycled and this material has been largely superseded by iron-gold alloys, the proportion of iron ranging from 0·02 atomic percent (Burns and Hurst, 1975) with the choice of iron content depending upon how the thermocouple is to be used (Berman and Kopp, 1971). However, at low temperatures, thermocouples containing iron are influenced by magnetic fields—an important consideration in thermomagnetic measurements. Alloys having a concentration of iron between 0.03 atomic per cent and 0.1 atomic per cent are least sensitive to magnetic field influence (Berman, 1972). Although at low temperatures the resistivities of thermocouples are less than those of resistance thermometers, the differential (back-to-back) connection of thermocouples enables small temperature differences to be measured accurately.

Thermocouples are used in the measurement of thermal conductivity from room temperature up to $\sim$ 1700 K. The two types of thermocouples employed depending upon the temperature range of measurement are platinum versus platinum/10% rhodium (Garber et al., 1963, Laubitz and McElroy, 1971) and chromel versus alumel (Laubitz and McElroy, 1971, Tye et al, 1972). Although the former thermocouple material is resistant to oxidation and can be used at a higher temperature, it does not offer any distinct advantage over chromel/alumel thermocouples in actual thermophysical measurement up to 1100 K by stationary (steady state) methods (Liermann, 1974).

Thermocouples have also been used to monitor temperature transients when measuring thermal diffusivity by the "flash method", but even intrinsic thermocouples have a finite response time (Sergeev, 1980) which will give rise to substantial errors in the data obtained from measurements on low diffusivity materials (Hennings and Parker, 1967, Walter et al, 1972).

Inaccuracies in measurement of thermal diffusivity by heat pulse techniques can result from the thermal inertia of the temperature sensor. Thermal inertia may be avoided by the use of superconducting bolometers or Tunnel junctions (Gutfeld, 1968). These sensors, however, are limited to a small temperature interval depending upon the transition temperature of the alloy used. The use of an avalanche sensor for detecting heat pulses in semiconductors has been proposed by Zylbersztejn (1967) but this device cannot be used for measurement on doped semiconductors. An ion-implanted resistor has been successfully used as a thermal transient sensor in the determination of the thermal diffusivity of silicon (Turkes, 1983).

Infrared detectors have found wide applications in the measurement of temperature transients; as there is no contact with the sample, there is no time delay. From room temperature upwards InSb or $Hg_x Cd_{1-x} Te$ detectors are used although during operation both types require some form of cooling. Above 500 K a lead sulphide detector can be used which does not require cooling.

2.2 Static (steady-state) methods

2.2.1 Introduction

Static or steady state methods provide the most accurate means of measuring thermal conductivity and can be divided into absolute and comparative methods. Absolute methods involve the direct measurement of the heat (usually in terms of electrical energy) passing through the sample. This method is employed at both low temperatures (axial method) or at high temperatures (radial method). At high temperatures it becomes increasingly difficult to allow for radiation heat losses and comparative methods may be more appropriate. In the comparative method the same flow of heat passes through the sample under test and a sample or samples of similar thermal conductivity placed in series with it. The flow of heat is then calculated from the temperature gradient over the standard sample(s).

2.2.2 Absolute axial-heat-flow method

A schematic diagram of the simplest steady-state experimental arrangement of the axial or longitudinal-heat-flow method is shown in Fig. 2.1. An

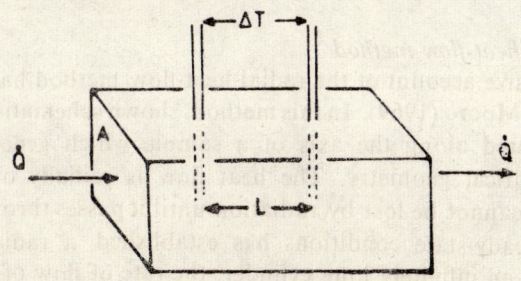

Fig. 2.1 Principle of the steady-state longitudinal heat flow method for measuring thermal conductivity.

electric heater supplies heat at rate $\dot{Q}$ to one end of a sample of uniform cross-section A and the heat is removed at the other end by a sink. Assuming that there are no heat losses or gains along the sample the mean thermal conductivity between the temperatures T_1 and T_2 is given by

$$\lambda = -\frac{\dot{Q}}{A}\frac{L}{\Delta T} \qquad (2.1)$$

where L is the separation between the thermocouples. Unless λ varies monotonically with temperature ΔT should be small in order to assign a meaningful value to λ.

In practice it is very difficult to ensure that all the heat supplied does go through the sample as heat exchanges occur between the sample surface and its surroundings and along thermocouple wires. In addition, it is often difficult and time consuming to ensure that the measurements are made under equilibrium conditions. Making measurements under vacuum ensures that any heat losses from the sample surface are radiative and at temperatures below about 100 K, careful design ensures that radiative losses are negligible compared to the heat conduction through the sample (Rosenberg, 1954, White and Woods, 1955, Slack, 1957). At temperatures from 100 K to room temperature, the radiation losses become significant and must be minimised or taken into account. The use of heated radiation shield around the sample and with a similar temperature gradient to that of the sample is effective in reducing radiation losses (Slack, 1961). Alternatively, the radiation losses can be estimated experimentally by changing the ratio L/A and taking them into account. As the temperature of measurement increases, so do the radiation heat losses. In measurements above room temperature the samples are usually enclosed in an evacuated furnace. Radiation and convection heat losses are minimised by surrounding the sample with an insulating filling of low thermal conductivity powder and locating a heated metal guard shield between the sample and the furnace wall. The temperature gradient in the shield is maintained the same as that along the sample. Nevertheless, maintaining a linear heat flow through the sample proves to be very difficult even with guard plates. Appropriate expressions for determining thermal conductivity by the axial heat-flow method, when heat losses occur through the filling, have been developed.

2.2.3 *Radial-heat-flow method*

A comprehensive account of the radial-heat-flow method has been given by McElroy and Moore (1969). In this method, shown schematically in Fig. 2.2, heat is generated along the axis of a sample which generally has right circular cylindrical geometry. The heat flow is radially outward, consequently, heat cannot be lost by radiation until it passes through the sample, and under steady-state conditions has established a radial temperature gradient. For an infinitely long cylinder, the rate of flow of heat per metre of sample length is then given by

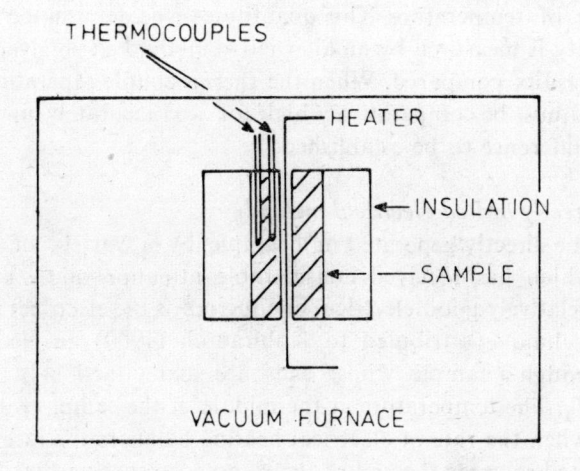

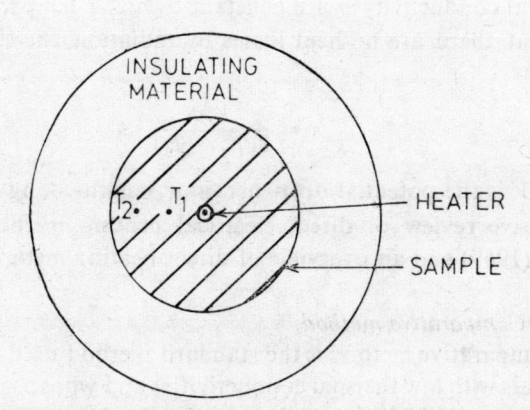

Fig. 2.2 Schematic diagram of radial flow measurement of thermal conductivity.

$$\dot{Q} = \frac{2\pi\lambda(T_2 - T_1)}{\ln(r_1/r_2)} \qquad (2.2)$$

where T_1 and T_2 are the temperatures at radii of r_1 and r_2, respectively. This method can be used on samples of finite length provided precautions are taken to prevent longitudinal heat flow. Errors in thermal conductivity due to end effects are negligible if the length-to-diameter ratio is greater than 12 (Kingery, 1954) and less than 5% if the ratio is greater than 4 (Slack and Glassbrenner, 1960). Radial-heat-flow techniques are generally used with large samples although, provided sufficient care is taken, measurements can be made on samples with dimensions of the order of a few centimetres (Glassbrenner, 1965). Longitudinal heat flow can be reduced by splitting the cylindrical samples into discs and providing a high thermal resistance at the disc interface. As the sample size is reduced, the main problem is the accurate measurement of the exact radial distance of each thermocouple from the centre. Any inaccuracy in this is a constant fraction which is independent of temperature. This quantity can be determined if the thermal conductivity is measured by an alternative method at a given temperature and the results compared. When the thermocouple separation is small the heat input must be comparatively high for an accurately measurable temperature difference to be established.

2.2.4 Directly heated electrical method

Heat can be directly generated in a sample by a variety of means, but a method which has received considerable attention and is appropriate for use with relatively good electrical conductors is the electrical method. Based upon a technique attributed to Kohlrausch (1900) an electric current is passed through a sample whose ends are maintained at a constant temperature T_1. The temperature at the middle of the sample will increase to a value T_2 when the rate of electrical heating balances the rate at which heat is conducted towards the end of the sample. Assuming that both λ and the electrical conductivity σ are constant over the temperature range involved and that there are no heat losses by radiation, the thermal conductivity is given by

$$\lambda = \frac{V^2 \sigma}{8(T_2 - T_1)} \qquad (2.3)$$

where V is the potential drop per unit length along the sample. A comprehensive review of direct electrical heating methods has been given by Flynn (1969) and an overview of direct heating methods by Taylor (1972).

2.2.5 Comparative method

The comparative method is the standard method used in the measurement of materials with low thermal conductivities and whose sample dimensions have a diameter-to-thickness ratio of greater than unity. This method has been used in its original or modified form to measure the thermal conductivities of a large number of semiconductor materials (Greico and Montgomery, 1952, McCarthy and Ballard, 1955, Stuckes and Chasmar, 1956, Bowers *et al*, 1959, Morris and Hust, 1961, and Morris and Martin, 1963).

In the comparative method the sample of unknown thermal conductivity λ_u is placed in series with one or more standard materials of known thermal conductivity λ_s. Ideally, the same flow of heat will pass through the whole column and under steady state conditions

$$\dot{Q} = \lambda_u A_u \left(\frac{\Delta T}{\Delta X}\right)_u = \lambda_s A_s \left(\frac{\Delta T}{\Delta X}\right)_s \qquad (2.4)$$

Hence
$$\lambda_u = \lambda_s \frac{(\Delta T/\Delta X)_s}{(\Delta T/\Delta X)_u} \qquad (2.5)$$

provided that $A_u = A_s$ (as is normally the case).

Here A's are the cross-sectional areas and ΔT the temperature difference between the two planar isothermal surfaces separated by a distance ΔX.

MEASUREMENT OF THERMAL CONDUCTIVITY 19

The thermal resistance of the sample and the standards will be directly proportional to the respective temperature gradients across each. The comparative method is of course based upon standard materials with accurately known thermal conductivity values. Armco iron has been used extensively but suffers from a high oxidation rate above 900 K. The apparatus used by Stuckes and Chasmar is shown in Fig. 2.3. They used stainless steel as a

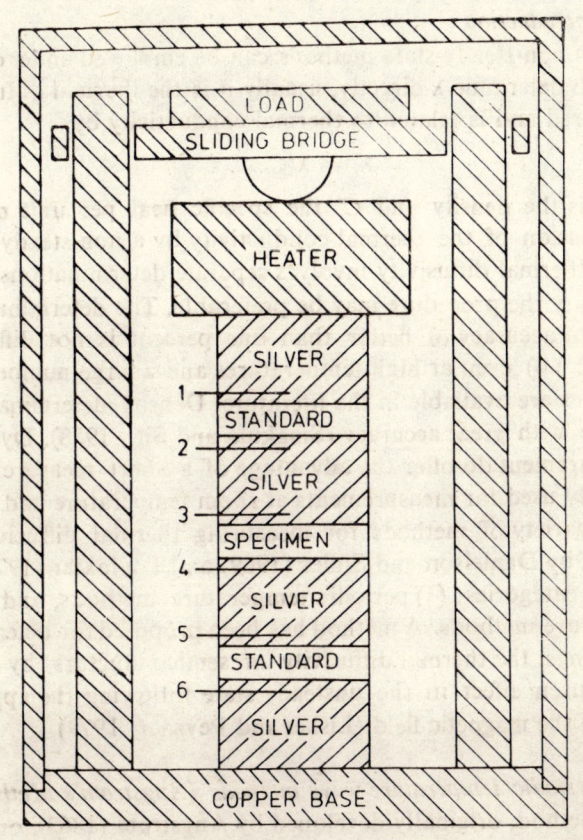

Fig. 2.3 Comparative method (Stuckes and Chasmar, 1956) (1–6) are Ni–Cr/Ni–Al thermocouples.

standard but recommended quartz for use with semiconductors of low thermal conductivity. The sample under measurement and two standard specimens are sandwiched between four blocks of silver and the whole stack compressed between a heater and the copper base. In order to ensure that there is a negligible temperature drop at each contact, thin layers of indium amalgam are employed. The heater element was constructed from nichrome wire wound on a stainless steel cylinder. A precise measurement of the thermocouple positions was not important since the temperature drop in the silver block was small compared to that in the standard or sample. In practice heat is exchanged between the sides of the sample, the standards and the surroundings, and to minimise these lateral losses the temperature drop across the standard should be small and approximately equal. This requires that $\lambda_u \simeq \lambda_s$. Lateral losses can also be reduced by the use of a guard ring and by polishing the components of the apparatus.

2.3 Non-steady-state (dynamic) methods

2.3.1 Introduction

Although non-steady-state methods can be employed under certain circumstances to determine λ directly, usually it is the thermal diffusivity λ' which is measured and is related to thermal conductivity by

$$\lambda' = \lambda/C'\rho \tag{2.6}$$

where ρ is the density and C' the specific heat per unit mass. Although determination of the thermal conductivity by a non-steady-state measurement of thermal diffusivity involves separate determinations of specific heat and density the procedure may be preferable. The determination of specific heat to an accuracy of better than one percent is not difficult (see sub-section 2.3.4) even at high temperatures and a large number of heat capacity values are available in the literature. Density determinations can also be made with great accuracy (Smakula and Sils, 1958). Dynamic methods of measurement do offer the advantage of a short measurement time and are widely used for measurements at room temperature and above.

The variety of methods for measuring thermal diffusivity have been reviewed by Danielson and Sidles (1969) and Toulokian (1973), and divided into two categories: (1) periodic temperature methods, and (2) transitory temperature methods. A method has been proposed for determining, among other things, the thermal diffusivity of semiconductors by measuring the Ettinghausen effect in the unsteady state following the application or removal of the magnetic field (Lisker and Pevzner, 1979).

2.3.2 Periodic temperature wave methods (Angstrom's Method)

In this method, originally developed by Angstrom (1863), one end of a long rod-shaped sample is heated periodically with a periodic time τ. The temperature wave is attenuated as it travels along the sample and as the wave has a finite velocity, there is a phase relationship between the signals measured with thermocouple at any two points along the sample's length. If the temperature wave is sinusoidal, the diffusivity is given by

$$\lambda' = \pi L^2/\tau\phi \ln \delta \tag{2.7}$$

where ϕ is the phase difference at two points along the sample, and separated by a distance L and δ is the amplitude ratio at the two points.

In order to satisfy the semi-infinite boundary conditions assumed in the derivation of Eq. 2.7, the temperature wave must decay to a negligible amount at the end of the sample. The temperature must also be strictly controlled, with the rise in temperature of the sample above that of the surroundings kept small. The validity of the method requires that the rate of heat loss from the surface of the sample be at all points proportional to the temperature difference between the sample and the surroundings. Also any change in ambient must be small compared with the amplitude of the sample's temperature wave. In addition problems are encountered in generating a symmetrical wave although the errors which arise from any asymmetry can usually be minimised by locating the thermocouples beyond a position from the source where the higher harmonics of the wave have been sufficiently attenuated, or by Fourier analysis of the two temperature variations.

A variety of methods have been employed in generating the temperature wave. Resistance heaters have been used under on-off conditions (Kanai and Nii, 1959) or with a sinusoidal current (Abeles *et al.*, 1960). Green and Cowles (1960) used the Peltier effect in a semiconductor thermocouple to produce the temperature wave. This has the advantage that cooling as well as heating is possible and that an almost symmetrical temperature wave can be obtained; but the method is difficult to use even at moderately high temperatures. Samples can also be heated using an electron beam and a comprehensive review of electron bombardment modulated heat input methods has been given by De Coninck and Peletsky (1984).

A temperature wave can be produced by focussing the light from a projection lamp onto the surface of the top end of the sample (Meddins and Parrott, 1969). A sinusoidal temperature variation is produced by rotating a rectangular paddle between the lamp and the sample. The ratio of the amplitudes at the top and base of the sample is dependent on the length of the sample and the period of the temperature wave. While the period must be sufficiently short to ensure semi-infinite boundary conditions, excessive reduction of its magnitude must be avoided since this may reduce the amplitude of the temperature variations below the level at which they can be measured with acceptable accuracy. If an approximate room temperature value of thermal diffusivity is known it is possible to obtain an indication of the ratio of the signals at the top and bottom of the sample for a given period and length of sample by using Eq. 2.7. The appropriate period for a 6 cm-long sample of germanium is 20 sec and that for 4 cm-long samples of lead telluride and silicon-germanium about 30–60 sec. Although an accuracy of better than 10% is claimed the mechanical method used in generating the sinusoidal temperature wave limits the technique to relatively long periods and consequently measurements were restricted to long samples of low diffusivity.

A rather more versatile apparatus based on Angstrom's method has been described by Savvides and Murray (1978) in which a stable sinusoidal temperature condition is produced in the heater by a low-frequency sinewave generator. A digital phase meter is used to measure the phase directly with a resolution of 0.1°. An accuracy of the thermal diffusivity measurement of $\pm 2\%$ is claimed. Block diagrams of the measuring circuit and of the operation of the sine wave generator are shown in Figs. 2.4 (a) and (b) respectively. The design of the sinewave generator was based on a stable clock (frequency) generator followed by dividers using digital integrated circuits to obtain the required stability and range. A triangular wave is generated with digital-to-analogue converter and shaped to give a sinewave. In the experimental arrangement the thermocouple voltage is selected by an L and N rotary switch of low thermal emf and fed into one of the amplifiers. Two electrical outputs are available; one is fed to the phase meter, the other to the x-t recorder.

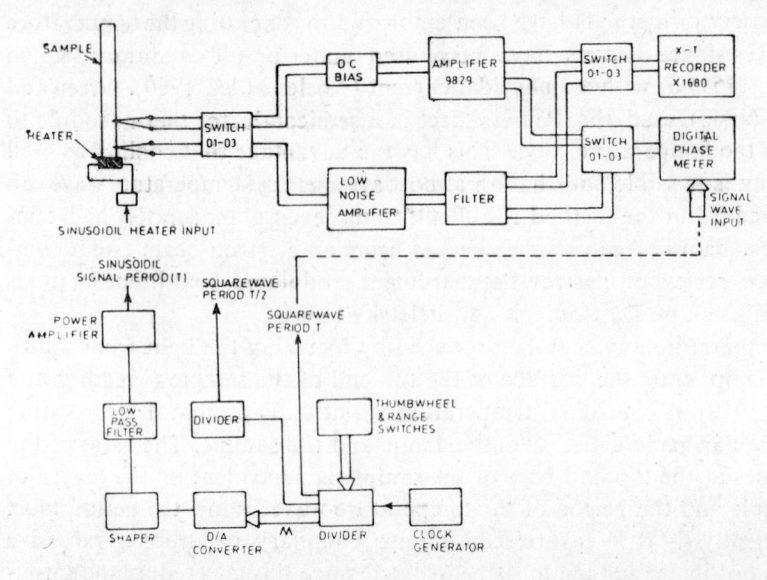

Fig. 2.4 (a) Block diagram of apparatus for measurement of thermal diffusivity featuring a low frequency sinewave generator and digital phasemeter.
(b) Block diagram of the sinewave generator (Savvides and Murray, 1978).

An apparatus that is capable of measuring a wide range of thermal diffusivity [from $10^{-2} - 1$ cm^2 S^{-1} and over a temperature range 80–500K] has been described by Vandersande and Pohl (1980). The main advantage of the apparatus, which is a modified Angstrom method, is its simplicity. All the electronic equipment is commercially available and the sample holder, [based on that used by Savvides and Murray, 1978] and the cryostat can easily be made in a short time. Accuracies in measurement of between $\pm 3\%$ to $\pm 7\%$ are claimed, depending upon the diffusivity of the sample and the temperature.

Temperatures are normally measured with thermocouples and it is the accuracy of measuring their separation which largely determines the overall accuracy of the thermal diffusivity measurement. Typically, the thermocouple separation can be measured to $\pm 0.5\%$ accuracy. Care should be taken to ensure that the thermocouples have a small thermal capacity to minimise heat losses along their length and to reduce thermal lag. Good contact must be made between the thermocouple and the sample and a variety of joining techniques have been developed. The choice of thermocouple material in the first instance depends upon the semiconductor being investigated as a number of them are reactive to normal thermocouple material. Copper reacts with silicon to form a low-melting-point copper-silicon alloy (Stapfer and Rouklove, 1972). A low-melting-point eutectic is also formed between platinum and silicon (Slack and Glassbrenner, 1960). Spot welding gives the best contact but when this proves difficult, satisfactory contacts can be made by wedging the thermocouples with the point of graphite pencil into small holes drilled ultrasonically into the sample.

2.3.3 *Transitory or pulse-heating methods*

In this method, originally described by Parker *et al* (1961), the front face of sample, usually in the form of a disc a few millimeters in diameter is irradiated with a short burst of radiant energy, the duration of which is short compared to the transit time of the pulse through the sample. The thermal diffusivity can then be obtained from the transient temperature history on the back face of the sample. The small sample-size requirement of this method is often an important consideration in a program of research when material may be at a premium. In addition the small sample size also means that it is possible to construct a system of small thermal inertia. This enables the ambient temperature to be changed quickly thus facilitating rapid measurements at different temperatures. Provided there are no heat losses from the sample the temperature at the near face will rise to a maximum value and remain constant. In principle, although any fractional rise in the temperature-rise-time curve can be used to compute the material's thermal diffusivity λ', the half-rise-time $t_{1/2}$ is most widely used where

$$\lambda' = \frac{0.1388\, L^2}{t_{1/2}} \qquad (2.8)$$

and L is the thickness of the sample.

In Parker *et al*'s original apparatus a xenon flash tube provided the heat pulse. The lamp's output is limited and some difficulty was encountered in achieving the required temperature rise in the backface surface temperature. A modified apparatus in which the light pulse from the flash lamp is fed to the sample by a fused quartz or sapphire light pipe, overcomes this drawback and measurements of thermal diffusivity using a flash lamp have been made up to 1600 K (Wood and Zoltan, 1984).

In later flash diffusivity apparatus, the flash tubes have largely been replaced by lasers (Righini and Cezairliyan, 1973).

It has been pointed out by Klemens (1984) that discrepancies can exist between thermal diffusivity values obtained from steady-state heat conductivities and those determined by flash diffusivity measurements. This arises because thermal conduction has a radiative component governed by an effective extinction length which is a function of photon frequency. There is a low-frequency spectral range for which photons are transmitted. Under transient conditions this part of the radiation plays no role in thermal diffusion and seems to reduce the apparent thermal diffusivity of the material.

Thermocouples have been extensively used to monitor the temperature on the rear face of the sample, and in order to minimise thermal lag the thermocouple wires are not joined at the end but separated by a small gap (1–2 mm) and pressed against the sample surface to complete the circuit. The signal voltage from the thermocouples is sensitive to contact resistance and it is often helpful to provide the surface of the sample with a conducting layer by rubbing heavily with a soft pencil. Precautions which should be taken during diffusivity measurements have been described in some detail by Taylor (1980). Apart from their thermal capacity interfering with the temperature measurement, thermocouples have a finite response time (Henning and Parker, 1967) which may lead to substantial errors in measurement, particularly in low diffusivity samples (Walter et al, 1972). In addition, semiconductors have low thermal diffusivities which necessitates the use of thin samples which are prone to fracture under the contacting pressure. The typical sample thickness for semiconductors such as Si-Ge alloys is around 1–2 mm.

Although the laser provides sufficient radiative energy to enable the sample to be located at some distance from the source and still give rise to a measurable temperature change on the rear surface, it does have a drawback in that the energy density profile across the laser beam is not uniform. The use of thermocouples which sense the temperature only over a small area of the sample surface can lead to significant errors. Radiation detectors which "look at" the average temperature are preferred for measurement from room temperature upwards. InSb and HgS detectors are available with response times of 1.5 µs and 5 µs respectively. The former is cooled with liquid nitrogen while the later is thermoelectrically cooled to −40°C. PbS detectors with a response time of 150 µs, can be used above 500 K.

The response of the thermocouple or infrared detector can be displayed on an oscilloscope and photographed or stored in a transient recorder and the $t_{1/2}$ value read off a fast chart recorder or computer analysed. A number of laboratories now undertake the measurement of thermal diffusivity on a commercial basis.

A laser flash system is shown in Fig. 2.5 (Rowe and Shukla, 1981). The

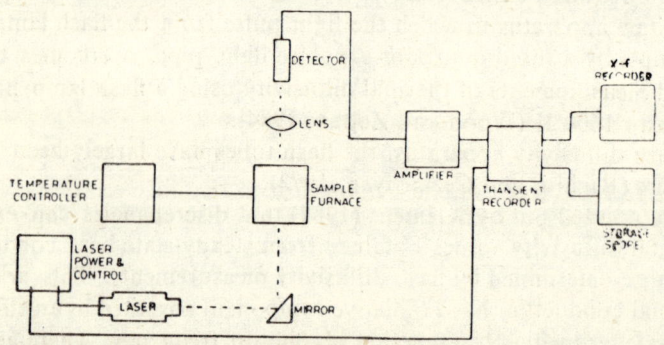

Fig. 2.5 Block diagram of laser flash diffusivity apparatus (Rowe and Shukla, 1981).

heat pulse is provided by a neodymium glass laser ($\lambda = 1.06$ μm), with a fixed pulse duration of 1 ms and a variable output up to a maximum of 20 J. In this particular apparatus the temperature history on the back-face of the sample is monitored with a thermocouple for measurements at room temperature and with a PbS detector at temperatures above 500K. The thermocouple is contacted to the sample using a technique described by Taylor (1975). The thermocouple wires of 0.12 mm-diameter chromel constantan are not joined at the end but separated by a gap of 1-2 mm and pressed against the near surface of the sample to complete the circuit. The signal voltage from the thermocouple is sensitive to the contact resistance of the thermocouple sample surface. Good reproducibility is obtained by heavily pencilling the face of the sample before making contact and by increasing the pressure on the thermocouple wires until the resistance of the thermocouple reaches a constant value. The PbS detector has a response time of 150 μs and is sensitive in the range 1-3 μm. The outputs from the thermocouple or the detector, typically 150 and 400 μV respectively, are amplified before passing into a transient recorder. A storage oscilloscope in parallel with the recorder provides a quick display of the signal which, if satisfactory, is recorded on a fast chart recorder. A transient of 1 ms duration corresponds to a chart length of 100 cm with an associated rise time of 20 cm. The $t_{1/2}$ value can be read to $\pm 1\%$. Accuracy and reproducibility of about $\pm 2\%$ at room temperature are claimed.

In practice the experimental arrangement will deviate in some degree from that dictated by the required boundary conditions and procedures have been described to allow for relatively long pulse times (Cape and Lehman, 1963, Taylor and Cape, 1964, Larson and Koyama, 1967) and for heat losses (Cape and Lehman, 1963, Cowan, 1963, Watt, 1966, Clark and Taylor, 1975). Errors due to a finite pulse time can be largely removed through a suitable choice of sample thickness. Estimates have also been made of the optimum sample size (Rykov, 1982). However, errors due to heat losses, particularly at high temperatures, cannot be avoided and they are usually taken into account using a practical solution due to Cowan (1963).

2.3.4 *Measurement of specific heat*

The specific heat of semiconductors can be accurately measured using a drop method (Ginnings *et al* 1950, Furakawa *et al*, 1956, and Maglic 1969). The sample is sealed in a container, heated to a known constant temperature and then dropped into a Bunsen ice calorimeter which measures the heat evolved by the sample and its container at nearly 0°C. Heat content of the empty container together with the small heat losses during the drop can be accounted for. The difference between the heat evolved in cooling from the given temperature in each case is a measure of the change in enthalpy of the sample between room temperature and 0°C. The procedure is repeated for a number of different temperatures and the heat capacity C_p determined using the expression

$$C_p = \left(\frac{\partial H}{\partial T}\right)_p \qquad (2.9)$$

In recent years established methods have been extended to facilitate the measurement of specific heat of new "exotic materials". A review of existing techniques and recent developments, such as the ac method, relaxation-time method, pulse method and laser-flash calorimetry has been given by Laxmikumar and Gopal (1981).

2.3.5 Thermal comparator

The thermal comparator method of obtaining a value of thermal conductivity was originally developed to provide a convenient method for arranging materials in order of increasing thermal conductivity (Powell, 1956 and 1957). Since its introduction in 1957 the method has demonstrated its versatility in the wide range of thermal conductivities that can be measured. Room-temperature measurements can be made quickly and on relatively small samples and, provided suitable precautions are taken, reasonably accurate thermal conductivity values can be obtained. The principle of operation of the thermal comparator depends upon monitoring the rate of cooling the tip of a heated probe when it is brought into contact with the surface of a material. A schematic diagram of the probe assembly is shown in Fig. 2.6. The probe utilizes a surface chromel-constantan thermocouple mounted at the tip of the probe which is differentially connected to the thermal reservoir. It quantitatively indicates the drop in temperature by the probe tip upon contact with the surface of a material. The observed reading is the differential emf between the exposed tip and a remote point on the top of the constantan block.

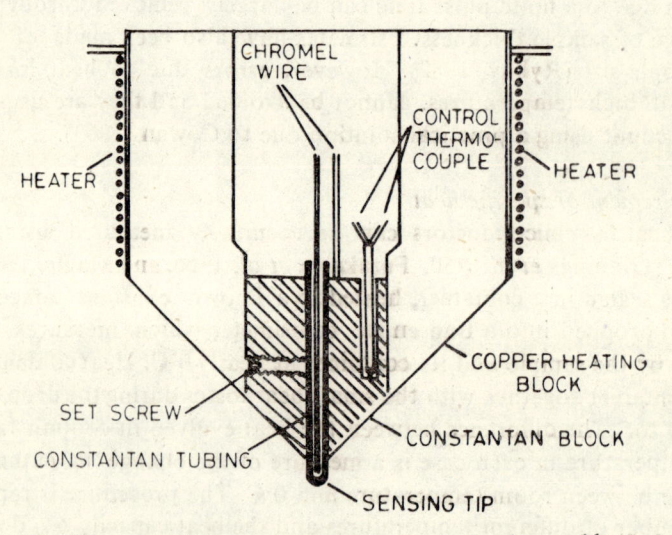

Fig. 2.6 Schematic diagram of comparator probe-head assembly (Lafayette Instrument Company).

A commercially available thermal comparator is shown in Fig 2.7. Experience has shown that proper performance of the device requires careful

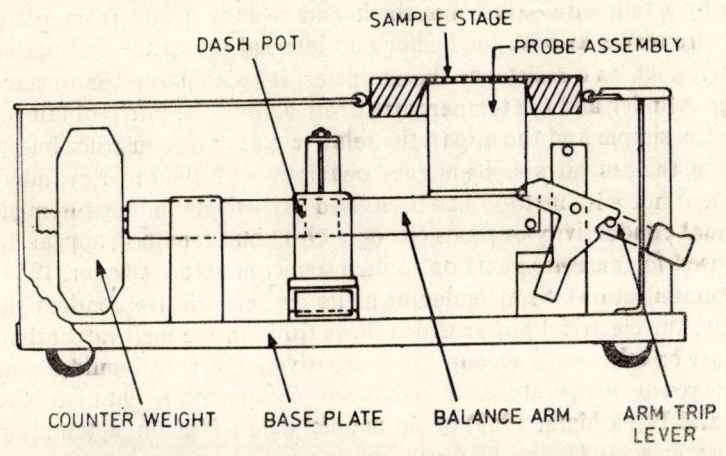

Fig. 2.7 Commercially available thermal comparator (Lafayette Instrument Company).

control of the pressure created by the probe and that the surface of the sample be clean and with a finish that can be routinely achieved by standard metallurgical practices. The probe is mounted on a counterbalanced arm supported by a precision bearing. The balance assembly is adjusted so that the load on the probe tip is 4g. A downward motion of the tip lever releases the balance arm, the probe moves upwards and touches the surface of the sample. The relationship between the thermal conductivities of various materials used as standards and the emf of the probe is provided in the form of a graph. For measurement of thermal conductivity, first one or two of these standards are measured and shown to agree with the graph, then the sample whose thermal conductivity is to be measured and which has nearly the same temperature and surface finish is contacted by the probe. The resulting emf output of the probe is referred to the curve and the thermal conductivity value is read.

2.4 Miscellaneous methods

2.4.1 *Semiconductor melts*

Molten semiconductors have received considerable attention in recent years particularly in the Soviet Union (Regel *et al.*, 1971), because of their potential use in thermoelectric applications. Thermal conductivity is an important parameter in assessing the "worth" of a thermoelectric material and measurements have been made on a large number of molten elemental and compound semiconductors (see Regel *et al*, 1971). Generally the same methods are employed in measurements on melts and on solid semiconductors. However, the methods employed often have modified names: axial heat flow is the horizontal flat plate cell and radial heat flow the coaxial cylinder cell (Tsederberg, 1965).

A method widely employed is the hot wire method (Ziebland, 1969). Essentially a modification of the coaxial cylinder method, the inner cylinder is replaced by a thin wire ~ 0.1 mm thick and usually made from platinum. The wire serves as both the heater and internal resistance. A suitable thermometer, such as a resistance thermometer, is located on the surface of the outer cylinder and the temperature drop across the melt obtained. The method is simple and the apparatus relatively easy to construct but it is crucial that the central wire be aligned coaxially with the outer cylinder.

A transient hot wire method has been used extensively in measurement of the thermal conductivity of fluids (Roder, 1981) but does not appear to have been used for measurements on molten semiconductors (Roder, 1984) (private communication). Semiconductor melts are electrically conducting, consequently, the electrical power which flows through the melt rather than the wire must be taken into account or alternatively the wire should be insulated. At room temperature, a polythene coating on a platinum wire (Nagasaka and Nagashima, 1983) or an anodic oxide film on a tantalum wire (Alloush *et al*, 1982) can be employed.

The modelling of semiconductor crystal growth from the melt requires information on their thermal properties in the molten state and in the solid state near the melting temperature. Lee and Taylor (1976) have described a technique for measuring the thermal diffusivity of a sample which is transparent to the laser pulse by sandwiching it between two opaque layers. Providing the semiconductor being investigated is sufficiently opaque at the two wavelengths of importance, namely those of the laser pulse and the detector, it is not necessary to provide a containment cell with opaque front or rear faces. Taylor and Lawrence (1984) measured the thermal diffusivity of a number of molten semiconductors. Fused silica cells were used to encapsulate $Hg_x Cd_{1-x} Te$ and quartz cells for Ge, PbTe and PbSnTe.

A few methods are suitable only for melt investigation. One example is a method based on the use of the thermal balance equation (Milvidskii and Eremeev, 1964) at the solid-melt boundary whereby the thermal conductivity is measured in the vicinity of the melting point directly in the process of crystal growth by the Czochralski method. Another is an optical method (Gustafsson, 1967) suitable for use with transparent melts with a change in the refraction coefficient depending upon the temperature increase.

In any event, the accurate measurement of the thermal conductivity of molten semiconductors is particularly difficult as there is the additional problem associated with taking into account convection heat losses. Containment of the sample also presents difficulties as most semiconductors are particularly reactive in the molten state. Consequently, several of the measuring arrangements can often only be used once.

2.4.2 *Thin films*

Semiconductor thin films are extensively employed in the microelectronics industry and are finding increasing applications in solar cell fabrication. Although effects due to thermal ageing must be of technological importance, the measurement of the thermal conductivity of semiconductor films has received little attention. A number of measurements have been made on non-semiconducting films, but the techniques employed are either questionable (Abrosimov 1969, Fraiman and Chudnovskii, 1971) or require special film geometry (Mathos and Ing, 1969).

A method for the accurate measurement of the thermal conductivity of thin samples of semiconductors with relatively high heat conductivities has been described by Savvides and Goldsmid (1972). In their method, which is an adaption of that first described by Harman (1958), the thin slice of semiconductor material B under examination (a good heat conductor) is sandwiched between two much thicker pieces of another material A which is a poor heat conductor as shown in Fig. 2.8 (a).

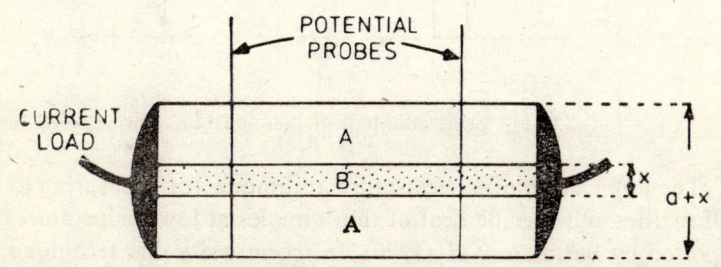

Fig. 2.8 (a) Three layer sandwich for thermal conductivity measurements (Savvides and Goldsmid, 1972).

When an electric current is passed through a conductor, the potential difference between any two points is made up of a voltage V_ρ due to the resistance and a thermoelectric voltage V_α which finds its origin in the temperature gradient resulting from the Peltier effect. The ratio between the two potentials is given by

$$\frac{V_\alpha}{V_\rho} = ZT = \frac{\alpha^2 T}{RK}$$

Z is the thermoelectric figure-of-merit, α is the Seebeck coefficient of the conductor (with respect to the potential probe material), T the absolute temperature, R the electrical resistance and K the thermal conductance between the two points. If, in the diagram, the electrical resistance of B is much larger than that of A, then the resistance R is essentially that of the two A's in parallel and is independent of x. Similarly, B does not make any reasonable contribution to the observed thermoelectric effects. However, it is assumed that both materials play a part in the conduction of heat. The thermal conductance of B can be obtained from measurements of the ratio of the thermoelectric potential difference to resistive potential difference with $x = 0$ (material B omitted) and $x = b$ (material B of thickness b included) and is given by

$$K_B = \frac{\alpha_A^2 T}{R_A} \left\{ \left(\frac{V_\rho}{V_\alpha}\right)_b - \left(\frac{V_\rho}{V_\alpha}\right)_0 \right\}$$

The use of inset potential probes results in a time lag before the thermoelectric potential difference changes after any change in the electric current (Penn, 1964). Consequently, V_ρ and V_α can readily be obtained by monitoring the overall potential difference between the potential probes on an $x - t$ recorder as the electric current is switched on and off. A schematic plot of potential difference against time is shown in Fig. 2.8(b).

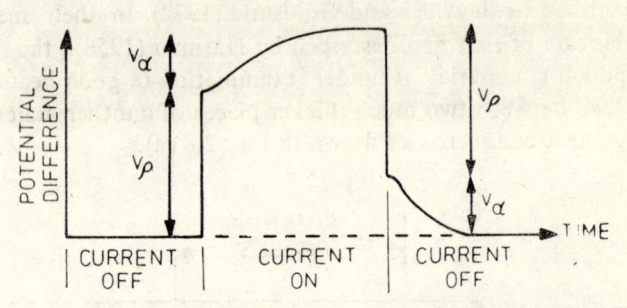

Fig. 2.8 (b) Schematic plot of potential difference against time.

The application of a heat-pulse technique for measuring the thermal diffusivities and specific heat of thin samples at low temperatures was first reported by Bertman et al. (1970). In recent years this technique has been developed and a number of refinements made to both the apparatus and the analysis (Filler et al, 1975, Cruz-Uribe and Trefny, 1979, Worthington et al, 1978, Cruz-Uribe and Trefny, 1982). In this method, a short pulse of heat (typically 1 μs) is introduced to the face end of a thin sample mounted on a substrate. The opposite end is anchored in good thermal contact to a low temperature support, the sample is otherwise isolated and in vacuum. The local time-dependent temperature response is monitored with a suitable thermocouple located some distance from the end of the sample. The thermal diffusivity and heat capacity per unit volume can then be obtained by comparing the observed temperature response which follows the initial heat input, with the solution of the one-dimensional diffusion equation, including the appropriate boundary conditions. An improved heat-pulse method described by Cruz-Uribe and Trefny (1982) has been used for measurements on amorphous semiconductors. Crystalline silicon substrates 0.05-0.08 mm thick were used and the temperature monitored using the resistance transitions of an evaporated superconducting film.

Two methods have been described by Nath and Chopra (1973 and 1974) for measuring the thermal conductivity of semiconductor films of thickness 500 A or more; a steady-state method (above room temperature), and a transient method (below room temperature). A modification of their multilayered method of measuring the thermal conductivity of thin semiconductor film has been described by Vigdorovich et al (1979), in which the specimen is heated without contacts by a high-frequency electromagnetic field.

2.4.3 *Photoacoustic effect*

The photoacoustic effect which was originally observed by Bell (1980) has attracted renewed interest following its development as a technique for the spectroscopic investigation of solid and semi-solid materials. The theory for the photoacoustic effect for solid samples has been presented by Rosencwaig and Gersho (1976). In the photoacoustic effect, periodically interrupted radiation is absorbed by a solid sample and gives rise to variations in the temperature of the sample. The variation in the pressure of the air in contact with the surface of the sample when confined in a closed volume (referred to as a photoacoustic cell) can be monitored with a sensitive microphone. A variety of methods based upon the photoacoustic effect for the measurement of thermal diffusivity have been described in the literature (Adams and Kirkbright, 1977, Ghizoni *et al*, 1978, Yasa and Amer, 1979, and Lepoutre *et al*, 1981).

A novel application of the technique which has been used in determining the thermal diffusivity of CdS is shown schematically in Fig. 2.9 (Cesar *et al*, 1983). Basically it uses the optical transparency of the sample to a given wavelength for generating, through a laterally modulated illumination, a periodic heat source at a point x_0 of the sample. The photoacoustic signal exhibits an experimental dependence on the product ax_0 between the illuminated region position x_0 and the thermal diffusion coefficient $a = (\pi t/\lambda')^{1/2}$.

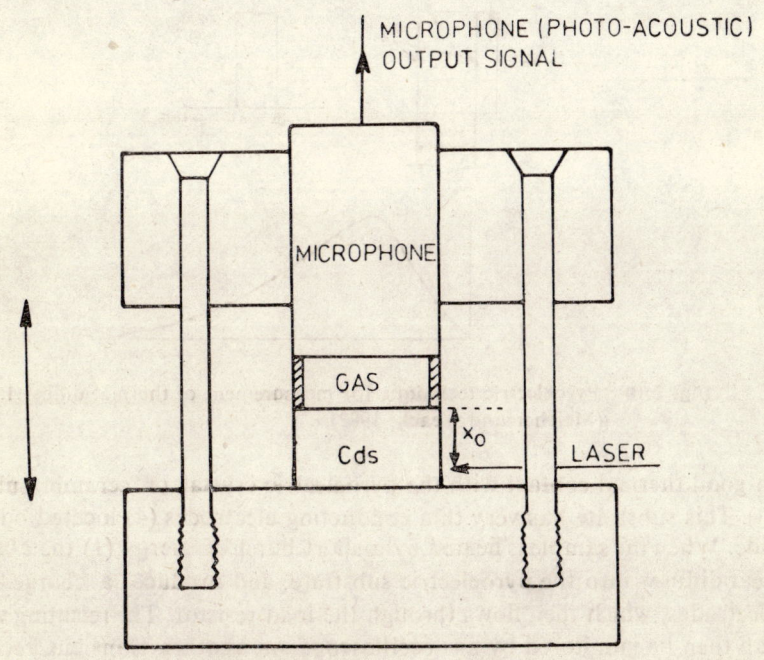

Fig. 2.9 Experimental setup used for determining the thermal diffusivity of CdS (Cesar *et al*, 1983).

This means that by measuring the acoustic signal, at a fixed modulation frequency t, as a function of x_0 the thermal diffusivity λ' is readily obtained as the function of x_0 in the exponential. In practice, this is achieved by moving the photoacoustic cell along by means of a micrometer positioner and varying the illumination region position x_0. The observed amplitude and phase of the acoustic signal are obtained as a function x_0.

An extension of the photoacoustic techniques is the photothermal laser probe (Williams, 1984). At frequencies near 1 GHz the acoustic wavelength in argon becomes half that of visible light, making possible Collinear Bragg scattering of laser light by the sound. This photo-optic interaction has been used to detect photothermally generated acoustic power at the surface of a sample when located in a cell filled with argon. The technique is sufficiently sensitive to changes in thermal properties to enable high resolution thermal images to be obtained of regions of a silicon sample which have been implanted with boron.

2.4.4 *Pyroelectric and AC colorimetric methods*

Pyroelectric sensors can be used to measure the thermal properties (Melcher and Yeack, 1982) in particular the thermal diffusivity of thin films, layers etc. The method is shown schematically in Fig. 2.10. The thin sample (2) is

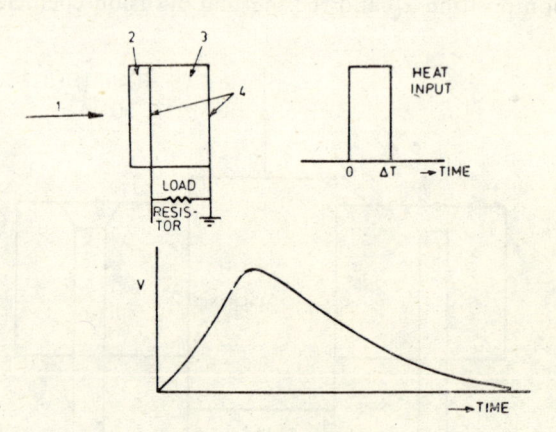

Fig. 2.10 Pyroelectric technique for measurement of thermal diffusivity (Melcher and Yeack, 1982).

in good thermal contact with the pyroelectric crystal or ceramic substrate (3). This substrate has very thin conducting electrodes (4) located on either side. When the sample is heated by a short burst of energy (1) the absorbed heat diffuses into the pyroelectric substrate, and produces a charge at the electrodes, which then flows through the load resistor. The resulting voltage can then be monitored by an oscilloscope or another transient recording system. The thermal diffusivity is then obtained from the shape of the voltage curve.

It has also been reported that when used as a thin-film calorimeter, a pyroelectric device has a time resolution of nanoseconds at a sensitivity of nano-calories (Coufal, 1984, 1985).

One problem encountered in the measurement of the thermal conductivity of thin films is the influence of the substrate. In order to obtain sufficient accuracy of measurement $\lambda_f d_f \gtrsim \lambda_s d_s$. Here, λ_f and d_f are the thermal conductivity and thickness of the film and λ_s and d_s refer to the substrate. In measurements on semiconductors where the thin film is mounted on a thin mica-on-glass substrate (Okun et al, 1969, Abrosimov et al, 1969, Borkov et al., 1975) $\lambda_s d_s \approx 6 \times 10^{-7}$ W K^{-1}. It follows that with semiconductor ($\lambda \sim 1$Wm^{-1}K^{-1}) films down to a thickness of 6×10^{-7} m can be investigated. This lower thickness limit has been substantially decreased by Volklein and Kessler (1984). In their method organic substrate foils ~ 40 nm thick with a thermal conductivity of $0.20 - 0.25$ Wm^{-1}K^{-1} are used, and this enables measurements to be made with sufficient accuracy on 10.4 nm-thick semiconductor films.

The thermal diffusivity of films in a direction parallel to the broad surface and when the sample thickness is less than the thermal skin depth has been measured using an ac calorimetric method (Hatta et al, 1986). Although measurements on semiconductors have not been reported, the technique is applicable to both high and low diffusivity material.

References

Abeles, B., Cody, G.D. and Beers, D.S. (1960), *J. Appl. Phys.* 31, 1985.

Abrosimov, V.M. (1969), *Sov. Phys.—Solid St*, 11, 2.

Abrosimov, V.M., Egorov, B.N., Lidorenko, N.S. and Rubasov, J.B. (1969), *Fiz. Tverd. Tela* 11, 530.

Adams, M.J. and Kirkbright, G.F. (1977), *The Analyst* 102, 678

Alloush, A., Gosney, W.B. and Wakeham, W.A. (1982), *Int. J. Thermop.* 3, 225

Angstrom, A.J. (1863), *Phil. Mag.* 25, 130.

Barber, C R. (1960), *Progr. Cryogen.* 2, 149.

Bell, A.G. (1980), *Am. J. Sci.* 20, 305.

Benedict, R.P. (1977), *Fundamentals of temperature, pressure and flow measurements*, second edition, Wiley, NY.

Berman, R. (1972), *IMCSI*, 4 prt 3, 1537.

Berman, R. (1976), *Thermal Conduction in Solids*, Clarendon Press, Oxford.

Berman, R. and Kopp, J. (1971), *J. Phys. F* 1, 457.

Bertman, B., Haberlein, D., Sandiford, D., Sheu, L. and Wagner, R (1970), *Cryogenics* 11, 326.

Billing, B F. and Quinn, T.J (ed.) (1975), *Temperature Measurement, Conf. Ser.* 26, Inst of Phys., London, p 144.

Borkov, Yu. A., Goltsman, B.M. and Sinenko, S.F. (1975), *Pribi. Tekh, Eksper,* 2, 230.

Bowers, R., Bauerle, J.E. and Cornish, A J. (1959), *J. Appl. Phys.* 30, 1050.

Broomfield, G.H. (1979), *The Metallurgist and Materials Technologist*, p. 201.

Burns, G.W. and Hurst, W.S. (1975), *Temperature* 75, 144.

Cape, J.A. and Lehman, G.W. (1963), *J. Appl. Phys.* 34, 1909.

Cesar, C.L., Vargas, H., Mendes Fihlo, J. and Miranda, L.C.M. (1983), *Appl. Phys. Lett*, 43, 555.

Clark, L.M. and Taylor, R.E. (1975), *J. Appl. Phys.* 46, 715.

Coufal, H. (1984), *Phys. Lett.* 44(1), 59.

Coufal, H. (1985), *Proc 5th Int. Symp. on Electrets*, Heidelberg.

Cowan, R.D. (1963), *J. Appl. Phys.* 34, 926.

Crovini, L. (1984), *High Temp, High Pressure* 17, 1.

Cruz-Uribe, A. and Trefny, J.U. (1979), *Cryogenics*. 19, 316

Cruz-uribe, A. and Trefny, J.U, (1982), *J. Phys. E*, Sci, Inst. 15, 1054.

Danielson, G.C. and Sidles, P.H. (1969), *Thermal Conductivity* (ed. R.P. Tye), Academic Press, London, p. 149.

De Coninck, R. and Peletsky, V E. (1984), *Compenduum of Thermophysical Property Meas. Methods*, Vol, 1 (ed. K. Moglic), plenum Press, p. 367.

Drabble, J.R. and Goldsmid, H J. (1961), *Thermal Conduction in Semiconductors*, Pergamon Press, Oxford.

Filler, R., Lindenfeld, P. and Deutsher, G. (1975), *Rev. Sci. Instrum.* 46, 439.

Flynn, D.R. (1969), *in Thermal Conductivity*, Vol. 1 (ed. R.P. Tye) Academic Press, New York, p. 241.

Fraiman, B.S. and Chudnovskii, A.F. (1971), *Fiz. Ter. Polyprov USSR* 5. 242.

Furakawa, G.T., Douglas, T.B., McCoskey, R.E. and Ginnings, D.C. (1956), *J. Res. NBS* 57, No. 2, 67.

Garber, M., Scott, B.W. and Blatt, F.J. (1963), *Phys. Rev.* 130, 288.

Ghizoni, G.C., Siquera, M.A.A., Vargas, H. and Miranda, L.C.M. (1978), *Appl. Phys. Lett.* 32, 554.

Ginnings, D.C. and Corrucini, R.J. (1947) *J. Res. NBS* 38, 593.

Ginnings, D.C., Douglas, T.B. and Balle, Anne, F. (1950), *J. NBS* 45, 23.

Glassbrenner, C.J. (1965), *Rev. Sci. Instrum.* 36, 984.

Green, A. and Cowles, E.J. (1960) *J. Sci. Instrum.* 37, 349.

Grieco, A. and Montgomery, H.C. (1952), *Phys. Rev.* 56, 570.

Gustafsson, S.E. (1967), *Z. Naturf.* 22 a, 1005.

Gutfeld, J.V. (1968), *In Physical Acoustics* (ed. P. Mason) Academic Press, London, Vol. 5, p. 233.

Hatta, I., Kato, R. and Maesono, A. (1986), 1st Asian Thermophysical Properties Conference, Beijing (China) p. 310 and 315.

Harman, T.C. (1958), *J. Appl. Phys.* 29, 1373.

Hennings, C.D. and Parker, R.A. (1967), *ASME J. Heat Transfer* 89C, 146.

Holland, M.G. and Rubin, L.G. (1962), *Rev. Sci. Instrum.* 33, 923.

Kanai, Y. and Nii, R. (1959), *J. Phys. Chem.* Solids, 8, 338.

Kingery, W.D. (1954), *J. Amer. Ceram. Soc.* 37, 88.

Klemens, P.G. (1984), *Proceedings of 9th European Thermophysical Properties Conference*, Manchester, Sept. 1984.

Kohlrausch, F. (1900), *Annln. Phys.* 1, 132.

Larson, K.B. and Koyama, K. (1967), *J. Appl. Phys.* 38, 465.

Laubitz, M.J. and McElroy, D.L. (1971), *Metrologiya* 7, No. 1, 1.

Laxmikumar, T. and Gopal, E.S.R. (1981), *J. Indian Inst. Sci.* 63, 277.

Lee, H.J. and Taylor, R.E. (1976), *Thermal Conductivity* 14, 423-434.

Lepoutre, F., Charpentier, P., Boccara, C. and Fourvier, D. (1981), *Topical Meeting on Photoacoustic Spectroscopy*, Berkley 1981, Paper MA3-1.

Liermann, J. (1974), *Bull. Inf. Sci. Tech. Commis. Energy. At* (Fr.) 197, 45.
Lindenfield, P. (1962), *Temperature: Its Measurement and Control in Science and Industry* 3, No. 1, 399, Reinhold, New York.
Llsker, I.S. and Pevzner, M.B. (1979), *J. Engg. Phys.* (USA) 36, part 3, 310.
Maglic, K.D. (1969), *MSc Thesis*, University of Wales.
Mathos, G.S. and Ing. P.W. (1969), *Pro. 8th Conf. on Ther. Cond.*, Plenum Press, New York.
McCarthy, K.A. and Ballard, S.S. (1955), *Phys. Rev.* 99, 1104.
McElroy, D.L. and Moore, J.P. (1969), *in Thermal Conductivity*, Vol. 1 (ed. R.P. Tye) Academic Press, p. 185.
Meddins, H.R. and Parrott, J.E. (1969), *Brit. J. Appl. Phys.* (J. Phys. D) Ser. 2, 691.
Melcher, R.L. and Yeack, C.E. (1982), IBM Tech. Dis. Bull, 25, 1, 69.
Milvidiskii, C.E. and Eremeev, V.V. (1964), *Fiz. Tverd. Tela.* 6, 1962.
Morris, R.G. aud Hust, I.G. (1961), *Phys. Rev.* 124, 1426.
Morris, R.G. and Martin, J.L (1963) *J. Appl. Phys* 34, 2388.
Nagasaka, H. and Nagashima, Y. (1983), *Thermal Conductivity*, (ed. J.B. Hust) Plenum Press, Vol 17, p. 307,
Naughton, M.J., Dickinson, S., Samaratungo, R.C., Brooks, J.S. and Martin, K.P. (1983), *Rev. Sci. Instrum*, 54, 11, 1529.
Nath, P. and Chopra, K.L. (1973), *Thin Solid Films* 18, 29.
Nath, P. and Chopra, K.L. (1974), *J. Appl. Phys.* (USA) 45, 4, 1923.
Okun, J.Z., Fraiman, B S., Cudnovskii, A.F. (1969), *Inz. fiz. Z.* 16, 334.
Parker, W.J., Jenkins, R.J., Butler, C.P. and Abbott, G.L. (1961), *J. Appl. Phys.* 32, 1679.
Parrott, J.E. and Stuckes, A.D. (1975), *Thermal Conductivity of Solids*, Pion, London.
Penn, A.W. (1964), *J. Scientific Instruments* 41, 626.
Powell, R W. (1956), *British Patent No.* 855, 658, Nov. 29,
Powell, R.W. (1957), *J. Sci. Instrum.* 34, 485.
Quinn, T.J. (1983), *Temperature*, Academic Press, London, New York.
Regel, A.R., Smirnov, I.A. and Shadrichev, E.V. (1971), *Phys. Stat. Solidi*, 5, 13.
Righini, F. and Cezairliyan. A. (1973), *High Temp.—High Pressures* 5, 481.
Roder, H.M. (1981), *J. of Res. of Nat. Bur. of Standards* 86, No. 5, 457.
Roder, H.M. (1984), *Private Communication.*
Rosenberg, H.M. (1954), *Proc. Phys. Soc.* A67, 837.
Rosencwaig, A. and Gersho, A, (1976), *J Appl. Phys.* 47, 64.
Rowe, D.M. and Shukla, V.S. (1981), *J. Appl. Phys.* 52, (12) 7421.
Rykov, E.S. (1982), *High Temp.* (USA) 20, 467.
Savvides, N. and Goldsmid, H.J. (1972), *J. Phys. E. Scient. Instrum.* 5, 553,
Savvides, N. and Murray, W. (1978), *J. Phys. E. Sci. Instrum.* 11, 941.
Sergeev, O.A. (1980), *Inzheneio—Fizicheskii Zhurnal* 39, No. 2, 306.
Slack, G.A. (1957), *Phys. Rev.* 105, 832.
Slack, G.A. (1961), *Phys Rev.* 122, 1451.
Slack, G.A. and Glassbrenner, C. (1960), *Phys. Rev.* 120, 782.
Smakula, A. and Sils, V. (1958), *Phys. Rev.* 99, 1744.
Stapfer, G. and Rouklave, P. (1972), *Proc of 7th IECEC*, San Diego, California, p. 149.
Stuckes, A.D. and Chasmar, R.P. (1956), *Report on the meeting on Semiconductors*, Physical Society; London, 119.

Taylor, R.E. (1975), *Rev. Int. Hautes Temp. Refr.* 12, 141.
Taylor, R.E. (1980), *J. Phys. E: Sci. Instrum.* 13, 1193.
Taylor, R.E. and Cape, J.A. (1964), *Appl. Phys. Lett.* 5, 212.
Taylor, R.E. and Lawrence, R. (1984), *9th European Thermophysical properties Conf.*, Manchester, Sept. 1984.
Toulokian, Y.S. et al. (1970), *Thermophysical Properties of Matter*, The TPRC Data Series Vol 1 and 2, IFI/Plenum Press, New York.
Toulokian, Y.S. (ed.) (1973), *Thermophysical Properties of Matter*, Vol. 10, Thermal Diffusivity, Plenum Press, N Y.
Tsederberg, N.V. (1965), *Thermal Conductivity of Gases and Liquids* (ed. R.D. Cess), Edward Arnold Ltd. London.
Turkes, P. (1983), *Phys. Stat. Solidi (a)* 75, 519.
Tye, R.P. (1970), *Rev. Int. Hautes Temper. et Refract.* 7, 380.
Tye, R.P., Hayden, R.W. and Spinney, S.C. (1972), *High Temp.—High Pressure* 4, No. 5, 503.
Tye, R.P. (ed.) (1969), *Thermal Conductivity*, Vol. 2, Ch. 4, Academic Press, London, New York.
Tye, R.P. and Hulstom, L.C. (1984), *Proceedings of 9th ETPC*, P.L 591 (ed. R. Taylor) Pion Ltd., London.
Vandersande, J.W. and Pohl, R.O. (1980), *Rev. Sci. Instrum.* 51 12), 1694.
Vigdorovich, V.N., Garanin, V.P. and Ukhlinov, G.A. (1979), *Ind. Lab (USA)/Zavad Lab* (USSR), Vol. 45, No. 5, 435.
Volklein, F. and Kessler, E. (1984), *Phys. Stat. Solidi (a)* 81, 585.
Walter, A.J., Dell, R M., Gilchrist, K.E. and Taylor, R.E. (1972), *High Pressure* 4, 439.
Watt, D.A. (1966), *Brit. J. Appl. Phys. Letters* 17, 231.
Williams, C.C. (1984), *Appl, Phys. Letters* 44, 1116.
White, G.K. (1969) in *Thermal Conductivity* (ed. R.P. Tye) Vol. 1, Academic Press, London, New York.
White, G.K. and Woods, S.B. (1955), *Can. J. Phys.* 33, 58.
Woodside, W. (1927), *ASTM—STP* 217 (American Society for Testing and Materials) p. 49.
Wood, C. and Zoltan, A. (1984), *Rev. Sci. Instrum.* (USA) 55, 2, 335.
Worthington, T., Lindenfeld, P. and Deutsher, G. (1978), *Phys. Rev. Letters* 41, 316.
Yasa, Z. and Amer, N.M. (1979), *Topical Meeting on Photo-acoustic Spectroscopy*, Ames 1979, Paper WA5—1.
Ziebland, H. (1969), *Thermal Conductivity* (ed. R.P. Tye), Vol. 2, Academic Press, Chapter 2.
Zylbersztejn, A. (1967), *Phys. Rev. Lett*, 19, 838.

Chapter III

Electrons and Holes in Semiconductors

3.1 Introduction

Electrons are important in the study of heat transport in semiconductors as they (along with holes) act as heat carriers and contribute to electronic thermal conductivity; they also act as scattering agents for phonons and thus influence the lattice thermal conductivity. A number of texts, for example Wilson (1953), Smith (1959), Drabble and Goldsmid (1961), Bube (1974), Kittel (1976) and Busch and Schade (1976) have discussed electronic behaviour in crystalline solids in considerable detail and we make no attempt to present every aspect of this topic. Rather, we consider only those aspects of electronic behaviour which have a direct bearing on heat transport in semiconductors. However, a brief description of the development of the subject is given as it is a prerequisite for understanding the basic concepts which lead to the formulation of heat transport equations and the methods employed in obtaining their solutions.

3.2 Electrons in solids

3.2.1 *Free electrons*

In many instances a simple model of electrons in a metal is found useful. This model referred to as a free electron model assumes non interacting (or free) electrons confined within a cubic box of side length L. The electron energy eigenvalues E_n and eigenfunctions ψ_n are given by

$$E_n = \frac{\hbar^2 \pi^2}{2mL^2}(n_x^2 + n_y^2 + n_z^2) \qquad (3.1)$$

$$\psi_{n_x, n_y, n_z} = \left(\frac{2}{L}\right)^{3/2} \sin(n_x \pi x/L) \sin(n_y \pi y/L) \sin(n_z \pi z/L) \qquad (3.2)$$

$\hbar = h/2\pi$, where h is Planck's constant, m is the electron mass and n_x, n_y, and n_z are positive integers. Each set (n_x, n_y, n_z) refers to an electron state which is in general degenerate. The ground state energy corresponds to $n_x = 1$, $n_y = 1$, $n_z = 1$ and for a cube of side 1 cm it would be almost zero for all practical purposes. The difference in energy between two consecutive levels is also vanishingly small and the energy levels appear to form an almost continuous (quasi-continuous) distribution. A density-of-states $D(E)dE$ can be defined as the number of such orbital states with energies between E and $E + dE$ and some simple considerations lead to:

$$D(E)dE = \frac{V}{4\pi^2} (2m/\hbar^2)^{3/2} E^{1/2} dE \qquad (3.3)$$

where V is the volume of the solid.

At a temperature $T = 0$ all states below a certain energy value E_F (referred to as the Fermi energy) are filled with electrons with each of the states accomodating two electrons of opposite spin; all states above E_F are empty.

3.2.2 *Periodic boundary conditions*

The model employed to describe electrons in metals (the Sommerfeld model) gives eigenfunctions in the form of standing waves. The expectation value of the momentum for this form of the wavefunction vanishes. In order to discuss the various transport phenomena a travelling wave representation is necessary. In the one-dimensional case the travelling wave associated with a free electron moving along the x-direction can be described as

$$\psi = e^{ikx} \qquad (3.4)$$

This lead to

$$E = \frac{\hbar^2}{2m} k^2 \qquad (3.5)$$

However, this travelling wave does not satisfy the boundary conditions appropriate to a box, i.e. $\psi = 0$ at $x = 0$ and $x = L$. In order to overcome this difficulty the one-dimensional solid under investigation is considered to be in the form of a wire and bent into a closed loop so that the point $x = 0$ coincides with the point $x = L$. The boundary conditions are then given by

$$\psi(0) = \psi(L)$$

and

$$\left(\frac{d\psi}{dx}\right)_{x=0} = \left(\frac{d\psi}{dx}\right)_{x=L}$$

In general, for one dimension, $\psi(x) = \psi(x + L)$. This gives

$$e^{ikL} = 1$$

or

$$k = 2\pi n/L \; (n = 0, \pm 1, \pm 2,...) \qquad (3.6)$$

3.2.3 *Periodic potential—origin of the energy bands*

Although the Sommerfeld model is able to explain a number of metallic properties, in many respects it is found inadequate. For example the electrons in a solid move in a periodic potential and this should be taken into consideration when formulating a theoretical model. Consider again a one-dimensional model with a periodic potential

$$V(x) \equiv V(x + a)$$

where a is the period of the lattice. The electron wavefunction obtained as a solution of the Schrodinger equation has the form

$$\psi_k(x) = e^{ikx} u_k(x) \tag{3.7}$$

where $\quad u_k(x) = u_k(x + a)$

The introduction of a periodic potential has important consequences. An arbitrary periodic potential gives rise to a series of allowed and forbidden bands. The origin of energy bands in solids can be discussed in different ways. For example it is well understood that in isolated atoms the electron energy levels are discrete and that when a number of atoms are brought together (to form a solid) the discrete levels broaden into bands. Bragg reflection is a feature of wave propagation in crystals and the inability of the electron with the appropriate energy to undergo a Bragg reflection supports the concept of forbidden energy gaps.

In the Bloch function $e^{ikx} u_k(x)$, k is not uniquely specified and if k is replaced by $k + (2\pi n/a)$ the form of the function is unaffected. On this basis it is possible to restrict the value of k to an interval of length $2\pi/a$, for example from $-\pi/a$ to $+\pi/a$. This range of k defines the first Brillouin zone or the fundamental domain of k.

The variation of electron energy with k is no longer given by Eq. (3.5). At the zone boundaries the energy levels are split and assuming that a small periodic potential (referring to the one-dimensional situation) can be treated as a perturbation, then at $k = n\pi/a$

$$E(k) = \frac{\hbar^2}{2m}\left(\frac{n\pi}{a}\right)^2 \pm |H'_n| \tag{3.8}$$

where $\quad H'_n = \dfrac{1}{a}\int \exp(-2\pi inx/a)\, V(x)\, dx.$

This leads to energy gap of $2 \mid H'_n \mid$ at $k = n\pi/a$. In this nearly-free-electron case the periodic potential gives rise to forbidden energy gaps in the otherwise continuous E–k curves and results in the formation of a series of allowed bands separated by forbidden regions or energy gaps.

These concepts are equally valid in a three-dimensional case where the Bloch functions are given by

$$\psi_k(\mathbf{r}) = \exp(i\mathbf{k}\cdot\mathbf{r})\, u_k(\mathbf{r}) \tag{3.9}$$

where $\quad u_k(\mathbf{r}) = u_k(\mathbf{r} + \mathbf{R}_n)$

and $\quad R_n = n_1 \mathbf{a}_1 + n_2 \mathbf{a}_2 + n_3 \mathbf{a}_3$

define the lattice translation vectors while $\mathbf{a}_1$, $\mathbf{a}_2$ and $\mathbf{a}_3$ are vectors which define the primitive cell. It is convenient to work in terms of reciprocal lattice vectors defined as

$$\mathbf{a}_i \cdot \mathbf{b}_j = 2\pi \delta_{ij} \qquad i, j = 1, 2, 3 \tag{3.10}$$

and expressed explicitly by

$$\mathbf{b}_1 = \frac{2\pi}{\Omega} \mathbf{a}_2 \times \mathbf{a}_3$$

$$\mathbf{b}_2 = \frac{2\pi}{\Omega} \mathbf{a}_3 \times \mathbf{a}_1$$

$$\mathbf{b}_3 = \frac{2\pi}{\Omega} \mathbf{a}_1 \times \mathbf{a}_2 \tag{3.11}$$

where $\Omega = \mathbf{a}_1 \cdot \mathbf{a}_2 \times \mathbf{a}_3$ is the volume of a unit cell. The first Brillouin zone in three dimensions can then be obtained as follows: draw the reciprocal lattice vectors joining the origin at $\mathbf{k} = 0$ to other reciprocal lattice points and construct planes that perpendicularly bisect these vectors. The smallest domain surrounding the origin and bound by these planes is the first Brillouin zone.

The volume of the unit cell in reciprocal space is given by
$$\mathbf{b}_1 \cdot \mathbf{b}_2 \times \mathbf{b}_3 = 8\pi^3/\Omega \text{ and } N = 1/\Omega$$
gives the number of unit cells per unit volume of the crystal. The number of allowed k-values within the first Brillouin zone is equal to N and these are distributed uniformly throughout the zone. The number of allowed k-values in a small element $d\mathbf{k}$ of $\mathbf{k}$ space is given by $(2/8\pi^3)\, d\mathbf{k}$ per unit volume of the crystal; allowance having been made for the fact that each k-value corresponds to two quantum states with opposite spin.

3.3 Metals, insulators and semiconductors

The band theory has successfully explained various electrical properties of

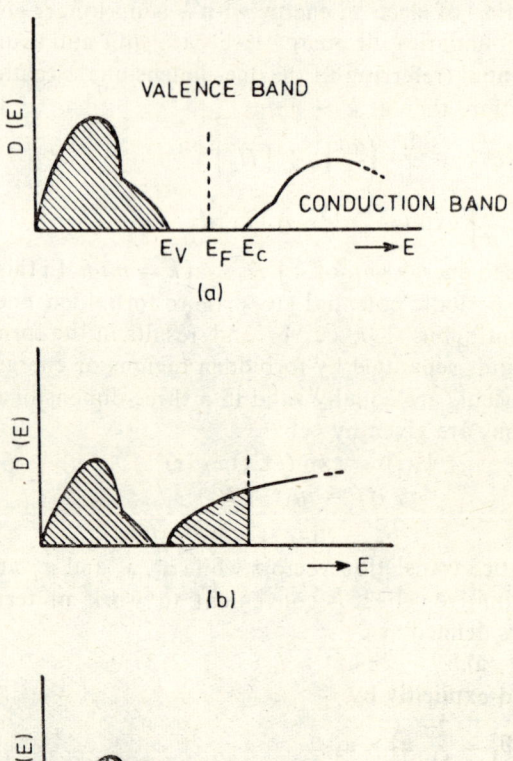

Fig. 3.1 Electronic density-of-states for (a) semiconductor or insulator, (b) a metal, and (c) a semi-metal.

ELECTRONS AND HOLES IN SEMICONDUCTORS 41

solids. In a lattice with one atom per unit cell a Brillouin zone can accomodate two electrons per atom. In real crystals there are several electrons per atom and the occupied states extend over several zones. In the simplest situation an even number of electrons per atom will lead to a completely filled zone or band and for an odd number of electrons per atom, the uppermost occupied band will be half-filled.

Materials which possess a completely filled uppermost band are likely to be insulators as transport of charge cannot take place in such bands. However, if some electrons can be thermally excited into the next higher band conduction will be possible. If the energy gap between the two bands is large compared to $k_B T$, the number of thermally excited electrons will be few. Such a material may exhibit semiconducting properties at high temperatures and will be an insulator at low temperatures. Metals are characterized by having the uppermost band partially filled. Typical density-of-states versus energy curves are shown in Fig. 3.1.

3.4 Electrons and holes

3.4.1 *Intrinsic conduction*

Consider a full band in a semiconductor (i.e. the valence band) with some of the electrons from the top of the band excited thermally across the energy gap to the next higher band (the conduction band). In the presence

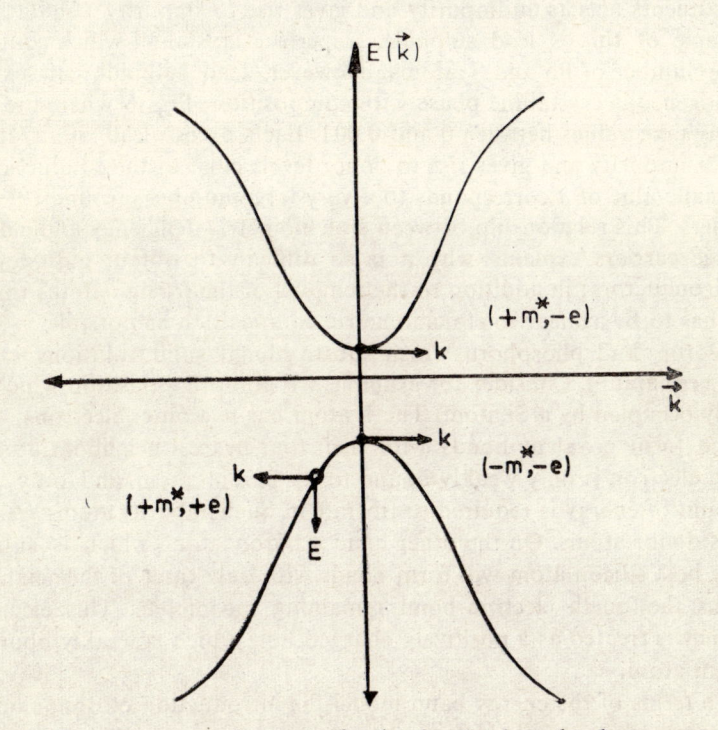

Fig. 3.2 Electrons at the conduction band and valence band edges and holes at the valence band edge.

of an external electric field, there will be a flow of charge in the valence band with sites vacated by the electrons moving in a direction which is opposite to that of the electrons in the conduction band. Such a vacant site in a band otherwise filled with electrons is called a hole and conduction in a nearly full band can be considered in terms of the motion of positively charged carriers. Electrons at the bottom of the conduction band have a positive effective mass and a negative charge while holes at the top of the valence band have a positive effective mass and a positive charge (Fig 3.2).

If a sufficient number of electrons is thermally excited across the energy gap into the conduction band the subsequent movement of electrons in the conduction band and holes in the valence band gives rise to electrical conduction. This type of conduction is called intrinsic conduction and in most semiconductors becomes significant only at high temperatures.

3.4.2 *Impurity or extrinsic conduction*

The addition of certain types of impurities and the introduction of imperfections into the host crystal lattice can drastically alter the electrical properties of the material (see, for example, Smith, 1959, and Kittel, 1976). For instance the addition of boron to silicon in the proportion of 1 to 10^5 increases the electrical conductivity by a factor of 10^3 at room temperature. In some compound semiconductors a stoichiometric deficiency in one of the constituents acts as an impurity and gives rise to impurity conduction. An example of this is lead sulphide, a perfect crystal of which contains the same number of Pb and S atoms. However, lead sulphide can exist as a homogeneous crystalline phase with composition $Pb_{1+\delta}S$ where the fraction δ may take values between 0 and 0.001. Each excess lead atom acts as a donor impurity and gives rise to donor levels close to the conduction band. A small value of δ corresponds to a very large number (around 10^{24} m^{-3}) of carriers. This relationship between stoichiometric deficiency and additional charge carriers explains why it is so difficult to obtain pure compound semiconductors; in addition to the removal of the foreign atoms the material has to be as near to stoichiometric composition as possible.

Boron and phosphorus form substitutional solid solutions in silicon and germanium. Consider for example a P atom in a position which is normally occupied by a Si atom. The P atom has five outer electrons, four of which form covalent bonds with their four nearest neighbour atoms. The extra electron is only weakly bound to its parent atom and only a small amount of energy is required to liberate it. Such impurity atoms are referred to as donor atoms. On the other hand a boron atom which is substituted for a host silicon atom will form bonds with only three of the neighbouring atoms, the fourth electron-bond remaining incomplete. This electron deficiency is treated as a positively charged hole which is weakly bound to its parent atom.

In terms of the energy band model the introduction of donor or acceptor impurities into the lattice introduces donor or acceptor energy levels close to the bottom of the conduction band and the top of the valence band, respectively. In Si and Ge with energy band gaps of 1.2 and 0.78 eV phosphorus donor levels are around 0.05 and 0.01 eV below the band edge. The ionization of donor impurities gives rise to negative charge carriers (electrons) in the conduction band while acceptor impurities give rise to positive charge carriers (holes) in the valence band. At low temperatures the carriers are "frozen" into the donor or acceptor levels. If the concentration of impurity atoms is very high, the wavefunctions of bound carriers (those in donor or acceptor levels) may overlap and form a band. This impurity band may become broad enough to overlap the conduction or valence bands. This situation is similar to that in a semimetal with a few electrons being available for conduction at all temperatures.

3.5 Energy bands in real crystals

In a simple theoretical model the extrema of the energy bands are assumed to lie at **k** = 0 and the equal energy surfaces are taken to be spherical with a scalar effective mass independent of **k**. Although in some cases the properties of real crystals can be qualitatively analysed within this framework, in general this simple model is inadequate. Consider the situation in germanium: the conduction band has several different extrema and the valence band has a maximum at **k** = 0. The minimum separation between the conduction and the valence bands is given by the difference in energy at the L points of the [111] zone face and the valence band extrema at the Γ point. The extrema of the two bands lie at different k values and this type of energy band gap is referred to as an indirect band gap. A second conduction band minimum occurs at the Γ point and the energy difference between the two extrema at Γ is called the direct band gap. In germanium the direct band gap is 0.16 eV larger than the indirect band gap.

In germanium the conduction band minima occur at the zone faces in the [111] directions. There are thus eight symmetry related half ellipsoids with long axes along the [111] directions centred on the midpoint of the zone face (Fig. 3.3). With a suitable choice of primitive cell in k-space these can be represented as four ellipsoids; the half ellipsoids on opposite faces can be joined by proper translations. In silicon the conduction band has six symmetry related minima along the [100] direction at points about eighty percent of the way to the zone boundary. The equal energy surfaces near the minima are ellipsoids of revolution with energy near the band edge given by

$$E(k) = \frac{\hbar^2}{2}\left[\frac{2k_T^2}{m_T^*} + \frac{k_L^2}{m_L^*}\right] \tag{3.12}$$

Here E is measured from the band extrema and T and L refer to the transverse and longitudinal mass parameters with respect to the principal axes of the equal energy ellipsoid. For silicon and germanium cyclotron resonance experiments lead to the following values of the effective masses at 4 K (m_0 being the free electron mass)

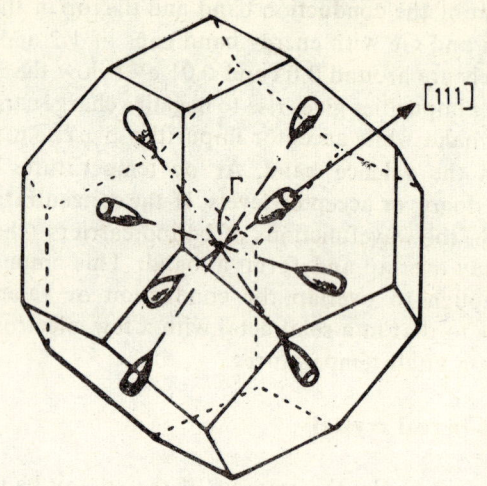

Fig. 3.3 The Brillouin zone and surfaces of constant energy near the conduction band in germanium.

$$m_T^* = 0.19\, m_0 \qquad m_L^* = 0.98\, m_0 \qquad \text{in silicon,}$$
and
$$m_T^* = .082\, m_0 \qquad m_L^* = 1.57\, m_0 \qquad \text{in germanium}$$

There are two degenerate valence band maxima at $\mathbf{k} = 0$ in both Si and Ge with effective masses $0.49\, m_0$ and $0.16\, m_0$ for silicon and $0.28\, m_0$ and $0.044\, m_0$ for germanium. There is a third band in both Si and Ge which is separated from the other two as a result of spin-orbit interaction.

For the electrons in the conduction band a density-of-states effective mass is defined which is given by

$$m_d^* = (m_T^{*2} m_L^*)^{1/3}. \tag{3.13}$$

The conductivity (or inertial) effective mass is given by

$$m_c^{*-1} = \frac{1}{3}(m_L^{*-1} + 2m_T^{*-1}) \tag{3.14}$$

For a multivalley semiconductor effective mass will be given by

$$m_d^* = N_v^{2/3}(m_L^* m_T^{*2})^{1/3} \tag{3.15}$$

N_v being the number of equivalent valleys.

3.6 The Fermi level

It is very convenient to express various electronic transport coefficients in terms of the Fermi level. In a metal at 0 K all energy states up to the Fermi level are filled with electrons while states above it are empty. At a non-zero temperature the energy states in a range of $\simeq 4k_B T$ about the Fermi level are partially filled. In a semiconductor, however, the situation is not that straightforward, although the possibility of a state being occupied depends upon the energy difference between the state and the Fermi level in exactly the same way as in metals.

A general model of a semiconductor consists of a conduction band and a valence band separated by an energy gap with the localized states corresponding to donor and acceptor impurities existing near the band edges (Fig. 3.4). Taking the zero of energy at the conduction band edge one defines

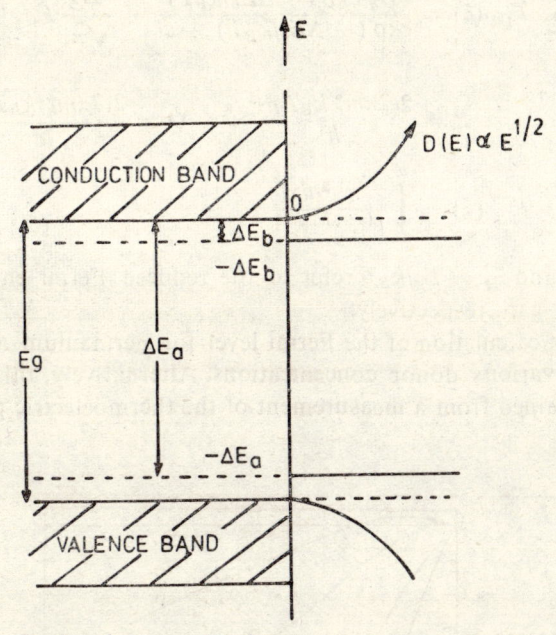

Fig. 3.4 Semiconductor model showing conduction and valence bands.

the energies corresponding to donor and acceptor levels as $-\Delta E_b$ and $-\Delta E_a$ respectively. The valence band edge lies at an energy $-E_g$.

The density-of-states $D(E)$ near the conduction band edge (for $E > 0$, where E is the energy of the charge carriers) and the valence band edge (for $E < -E_g$) is assumed to vary parabolically with E and is given by

$$D(E) = \frac{4\pi}{h^3}(2m_e^*)^{3/2} E^{1/2} \qquad \text{for } E > 0$$

$$D(E) = \frac{4\pi}{h^3}(2m_h^*)^{3/2} (|E| - E_g)^{1/2} \qquad \text{for } E < -E_g \qquad (3.16)$$

where m_e^* and m_h^* are the density-of-states effective masses for electrons and holes, respectively. In the gap region $D(E)$ at each impurity level can be written in terms of δ-functions

$$D(E) = D_a \delta(E + \Delta E_a) + D_b \delta(E + \Delta E_b) \qquad (3.17)$$

D_a and D_b are the number of states per unit volume in acceptor and donor levels.

In extrinsic semiconductors the position of the Fermi level depends upon the concentration of electrons in the conduction band (or holes in the valence band) and upon temperature. In the intrinsic range with equal concentrations of electrons and holes the Fermi level lies at the middle of the gap. The position of the Fermi level as a function of carrier concentration and temperature can be obtained from the following equation (Hutner *et al*, 1950)

$$\frac{2P}{\sqrt{\pi}} F_{1/2}(\xi) = \frac{D_b \exp(-\Delta E_b/k_B T)}{\exp(-\Delta E_b/k_B T) + e^\xi} + \frac{2Q}{\sqrt{\pi}} F_{1/2}(-\xi - \xi_g)$$

(3.18)

Here
$$P = \frac{2(2\pi m_e^* k_B T)^{3/2}}{h^3}; \quad Q = \frac{2(2\pi m_h^* k_B T)^{3/2}}{h^3}$$

and
$$F_{1/2}(\xi) = \int_0^\infty \frac{x^{1/2}\, dx}{e^{(x-\xi)} + 1}$$

$\xi = E_F/k_B T$ and $\xi_g = E_g/k_B T$ refer to the reduced Fermi energy and reduced energy gap, respectively.

Results of calculation of the Fermi level for germanium are shown in Fig. 3.5 for various donor concentrations. Alternatively, this information could be obtained from a measurement of the thermoelectric power (Fistul, 1969).

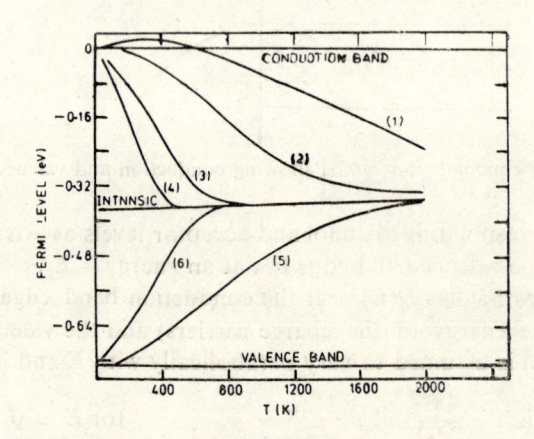

Fig. 3.5 Fermi level as a function of temperature and doping level in germanium; a two band model with a donor or acceptor level (after Hutner *et al.*, 1950).
Density-of-states values (m^{-3})
(1) $D_b = 10^{26}$ (2) $D_b = 10^{25}$ (3) $D_b = 10^{23}$
(4) $D_b = 10^{21}$ (5) $D_a = 10^{25}$ (6) $D_a = 10^{23}$
$\Delta E_b = 0.005$ eV below the conduction band,
$\Delta E_a = 0.01$ eV above the valence band
$E_g = 0.75$ eV; $m_h^*/m_e^* = 1.2$

3.7 Electrons as carriers of charge and heat

Metals in general possess high electrical and thermal conductivities and in simple metals the ratio of the two conductivities at a fixed temperature is very nearly a constant (Wiedemann-Franz-Lorenz law). On the other hand non-metallic solids are in general poor conductors of electricity and heat. However, this is not universally true and a number of non-metallic solids are very good conductors of heat (see Chapters I and XI).

In heavily doped semiconductors it is possible to attain electron (hole) concentrations in excess of 10^{26} m^{-3}. The charge carriers in these materials make a significant contribution to the thermal conductivity and in a number of materials this contribution may be as high as 25—30 per cent at room temperature.

Drude (1900, 1902) obtained an expression for electrical conductivity. He assumed the electrons to move freely in a solid with a mean-free-path l (the average distance travelled by an electron before suffering a collision). Under the influence of an external electric field, the electrons are accelerated. The existence of a frictional force of some kind is necessary which along with the external field gives rise to a steady flow of electrons. In general the frictional force is provided by the interaction between the electrons and the vibrations of the lattice, and a characteristic time τ related to the mean-free-path can be defined. In a simple analysis τ is assumed to be independent of electron energy and of its direction of motion.

These simple considerations leads to the following expression for electrical conductivity

$$\sigma = ne^2 \tau/m \tag{3.19}$$

Here n is the number of electrons per unit volume. In metals $n \sim 10^{28}$ m^{-3}, and τ should be around 10^{-14} sec in order to obtain an agreement with the observed electrical conductivity at room temperature.

Somewhat similar considerations apply to the flow of electrons under the influence of a temperature gradient. For one type of carriers (electron or hole) a polar contribution to thermal conductivity arises which increases with an increase in carrier concentration. In doped semiconductors the electronic contribution to thermal conductivity is of considerable importance and its calculation will form the subject matter of Ch. IV. The following section gives a brief account of the general transport equations and aspects of irreversible thermodynamics which are relevant to transport theory. Other sections deal with the electron Boltzmann equation and its solution.

3.8 Transport processes in solids

In a solid the transport processes include the flow of charge or energy or both. These flows arise due to external fields such as electric field, magnetic field and temperature gradient which are sometimes referred to as "forces". In general, any force can give rise to any flow and the relationships between the various flows and fields define the transport coefficients in terms of parameters that are characteristic of the electrons and phonons in the solid. If the departure from equilibrium is assumed to be small a linear relationship can be written between the flows F_i and the forces X_i as follows (Callen, 1948)

$$F_i = \sum_{k=1}^{n} L_{ik} X_k \qquad (3.20)$$

If a suitable choice is made for the flows and forces, the following indentities hold

$$L_{ik} = L_{kl} \qquad (i, k = 1, 2, ..., n) \qquad (3.21)$$

When a magnetic field **B** acts on the system, the coefficients L_{ik} depend upon **B** and

$$L_{ik}(\mathbf{B}) = L_{ki}(-\mathbf{B}) \qquad (i, k = 1, 2, ..., n) \qquad (3.22)$$

These equations are called the Onsager relationships. With a suitable choice of the generalized "forces" and "flows" the rate of entropy (S) production in an irreversible process can be written as

$$\frac{dS}{dt} = \sum_{i=1}^{n} F_i X_i \qquad (3.23)$$

with F_i and X_i related through Eq. 3.20.

Simple considerations regarding the flow of electrons and of energy in a solid lead to a suitable choice of "force" and "flow". Consider a solid in contact with two reservoirs, one of energy and the other of electrons, so that in the steady state a steady flow is maintained through the solid and constant differences in electrochemical potential $\bar{\mu}$ and temperature T between two points are maintained. It is not difficult to show (Appendix B) that, with grad $\frac{1}{T}$ and grad $\frac{\bar{\mu}}{T}$ as the choice of forces, the components of flows $-\mathbf{J}$ and $\mathbf{W}$ can be expressed as

$$-J_i = \sum_{k=1}^{k=3} L_{ik}^{(1)} \frac{\partial}{\partial x_k}\left(\frac{\bar{\mu}}{T}\right) + \sum_{k=1}^{k=3} L_{ik}^{(2)} \frac{\partial}{\partial x_k}\left(\frac{1}{T}\right)$$

$$W_i = \sum_{k=1}^{k=3} L_{ik}^{(3)} \frac{\partial}{\partial x_k}\left(\frac{\bar{\mu}}{T}\right) + \sum_{k=1}^{k=3} L_{ik}^{(4)} \frac{\partial}{\partial x_k}\left(\frac{1}{T}\right) \qquad (3.24)$$

The total energy flow can be written as a sum of $\mathbf{W}_e$ and $\mathbf{W}_p$ where e and p refer to the electron and phonon systems, respectively, so that

$$L^{(3)} = L_e^{(3)} + L_p^{(3)}.$$

In general, no cross relationships exist between $L^{(2)}$ and $L_p^{(3)}$ (particle flow and energy flow in the electron system). However, $L_p^{(3)}$ which defines the energy flow in the phonon system as a result of spatial variations of $\bar{\mu}$, is determined by the strength of interaction between electrons and phonons and gives rise to the so-called phonon-drag effect (see Chapter IV, Appendix A).

The set of coefficients L_{ik} provide a complete description of the transport properties of the solid. The number of independent coefficients is reduced by taking the crystal symmetry into account. To describe various physical situations, Eq. (3.24) can be expressed in a slightly different form where electric current-density (**i**) and thermal current density (**w**) are written in terms of grad $(\bar{\mu})$ and grad (T) (see Drabble and Goldsmid, 1961). The observable quantities—the electrical and thermal conductivities—and the Seebeck and Peltier coefficients make their appearance and are related to the coefficients L_{ik}.

One of these equations which describes thermal current density **W**, under conditions of zero electric current **i**, gives the familiar relation $\mathbf{W} = -\lambda$ grad T, which defines thermal conductivity λ.

3 9 Electron Boltzmann equation and its solution

3 9.1 *Distribution function for electrons*
Statistical methods are usually employed while dealing with situations involving a large number of particles. A probability distribution function $f_n(\mathbf{k}, \mathbf{r}, t)$ which describes the occupancy of allowed energy states, is introduced. The most probable number of electrons in the volume element $d\mathbf{k}$ of **k** space at time t is given by

$$\frac{2}{8\pi^3} f_n(\mathbf{k}, \mathbf{r}, t)\, d\mathbf{k}$$

n is the band index and will normally be omitted from future equations.

3.9.2 *The Boltzmann equation*
The distribution in the neighbourhood of **r** may change due to a number of mechanisms.

(i) An external force **F** may bring about changes in f. The vector **k** will change at rate

$$\dot{\mathbf{k}} = \frac{\mathbf{F}}{\hbar}$$

while the distribution function changes at the rate

$$\left(\frac{\partial f}{\partial t}\right)_{\text{field}} = -\frac{d\mathbf{k}}{dt} \cdot \nabla_{\mathbf{k}} f = -\frac{1}{\hbar}\, \mathbf{F} \cdot \nabla_{\mathbf{k}} f \qquad (3.25)$$

(ii) Electrons from neighbouring regions may enter into regions near **r** while others may leave as a result of their spatial velocities $\mathbf{V}_k$ and

$$\left(\frac{\partial f}{\partial t}\right)_{\text{diffusion}} = -\frac{d\mathbf{r}}{dt} \cdot \nabla_{\mathbf{r}} f = -\mathbf{V}_k \cdot \nabla_{\mathbf{r}} f \qquad (3.26)$$

(iii) Carriers jump from one state to another due to a variety of processes: interaction with one another, with phonons, or with defects, etc. The basic problem is to obtain $\left(\frac{\partial f}{\partial t}\right)_{\text{coll}}$ (the subscript denoting collision) as this expression depends upon the characteristic properties of a material. Let $Q_k^{k'}$ be the probability per unit time for the scattering of carriers from state **k** to **k'**. The quantity $\left(\frac{\partial f}{\partial t}\right)_{coll}$ can then be written as the difference between the rate at which electrons enter the state **k** and the rate at which they are lost from it, i.e.

$$\left(\frac{\partial f}{\partial t}\right)_{coll} = \iint \left\{ Q_{k'}^{k} f(k') [1 - f(k)] - Q_k^{k'} f(k) [1 - f(k')] \right\} dk' \quad (3.27)$$

The scattering rate from state **k** to state **k'** is not governed solely by the intrinsic transition probability $Q_k^{k'}$; one must take into consideration the factor $f_k(1 - f_{k'})$ for the chance that initially the state **k** is occupied and the state **k'** is empty. According to the principle of microscopic reversibility $Q_k^{k'} = Q_{k'}^{k}$. Since the equilibrium distribution function depends only upon energy ($f_k^0 = f_{k'}^0$), one obtains

$$\left(\frac{\partial f_k}{\partial t}\right)_{coll} = \int \left\{ [f(k') - f^0(k')] - [f(k) - f^0(k)] \right\} Q_k^{k'} dk' \quad (3.28)$$

In general the departure of the distribution function from the equilibrium value $f_0(E)$ is described by

$$f(\mathbf{k}, \mathbf{r}) = f_0(E) + f'(\mathbf{k}, \mathbf{r}) \quad (3.29)$$

It is convenient to introduce a function $\phi(\mathbf{k}, \mathbf{r})$, where

$$f'(\mathbf{k}, \mathbf{r}) = -\phi(\mathbf{k}, \mathbf{r}) \left(\frac{\partial f_0}{\partial E}\right) \quad (3.30)$$

and it is now possible to write an expression for $\left(\frac{\partial f}{\partial t}\right)_{coll}$ in terms of ϕ, which is a measure of the deviation of the distribution from its equilibrium value and is given by

$$\left(\frac{\partial f}{\partial t}\right)_{coll} = \frac{1}{k_B T} \int Q_k^{k'} f_0(E) [1 - f_0(E')] [\phi(k') - \phi(k)] dk' \quad (3.31)$$

The total rate of change of the distribution function is now

$$\left(\frac{\partial f}{\partial t}\right) = \left(\frac{\partial f}{\partial t}\right)_{field} + \left(\frac{\partial f}{\partial t}\right)_{diff} + \left(\frac{\partial f}{\partial t}\right)_{coll} \quad (3.32)$$

In the steady state this must equal zero. Substituting for the various rates one obtains

$$-\mathbf{V}_k \cdot \nabla_r f - \frac{1}{\hbar} \mathbf{F} \cdot \nabla_k f = -\left(\frac{\partial f}{\partial t}\right)_{coll} \quad (3.33)$$

This is the Boltzmann equation in its general form. The basic problem is to solve this equation for f_k; the flow of electrons and that of energy can then be obtained from

ELECTRONS AND HOLES IN SEMICONDUCTORS

$$-\mathbf{J} = -\frac{1}{4\pi^3}\int \mathbf{V_k}\, f_\mathbf{k}\, d\mathbf{k}$$

and
$$W = \frac{1}{4\pi^3}\int E_\mathbf{k} \mathbf{V_k}\, f_\mathbf{k}\, d\mathbf{k} \qquad (3.34)$$

3.9.3 Solving the Boltzmann equation

Obtaining a solution of the Boltzmann equation is difficult owing to the complexity of the collision term. In the variational approach one writes the equation in a canonical form and employs a general variational principle in trying to obtain the solution. This subject has been dealt with in considerable detail by Ziman (1960). The procedure adopted in solving the electron Boltzmann equation is similar to that for the phonon Boltzmann equation; some simple aspects of this approach will be described in Chap. VI. Some aspects of the variational principle and its use in solving the Boltzmann equation are described in the appendix. For the present, our discussion shall be limited to the relaxation-time approach.

3.9.4 Relaxation time approach

It is possible to obtain a solution of the electron Boltzmann equation if one assumes the existence of a relaxation time τ defined by

$$\left(\frac{\partial f}{\partial t}\right)_{\text{coll}} = -\frac{f'(\mathbf{k},\mathbf{r})}{\tau} \qquad (3.35)$$

The Boltzmann equation now reads as

$$e\vec{\mathcal{E}}\cdot\frac{1}{\hbar}\nabla_\mathbf{k} f + \mathbf{V_k}\cdot\nabla_\mathbf{r} f = \frac{\phi}{\tau}\frac{\partial f_0}{\partial E} \qquad (3.36)$$

with the electric field $\vec{\mathcal{E}}$ supplying the external force. Equation 3.36 can be simplified using the following relationships:

$$\nabla_r f \simeq \nabla_r f_0 \qquad (3.37)$$

$$f_0 = [\exp(E-E_F)k_B T + 1]^{-1}$$

and
$$\nabla_k f \simeq \frac{\partial f_0}{\partial E}\hbar \mathbf{V} \qquad (3.38)$$

One can now write

$$\nabla_r f \simeq -\left[\frac{E-E_F}{T}\nabla_r T + \nabla_r E_F\right]\frac{\partial f_0}{\partial E} \qquad (3.39)$$

Using these simplifications, Eq. 3.36 is reduced to

$$e\vec{\mathcal{E}}\cdot\mathbf{V}\frac{\partial f_0}{\partial E} - (E-E_F)\frac{\partial f_0}{\partial E}\mathbf{V}\cdot\nabla_r \ln T - \nabla_r E_F \frac{\partial f_0}{\partial E}\cdot\mathbf{V} = \frac{\phi}{\tau}\frac{\partial f_0}{\partial E} \qquad (3.40)$$

ϕ can be expressed in terms of generalized "forces" $\nabla(E_F/T)$ and grad $(1/T)$; E_F is expressed in terms of the gradient of an external electrostatic potential ψ

$$\nabla E_F = -e\,\nabla\psi = e\vec{\mathcal{E}} \qquad (3.41)$$

$$\phi = -T\tau\,\mathbf{V}.[\mathbf{\nabla}(E_F/T) - E\,\mathbf{\nabla}(1/T)] \tag{3.42}$$

and the electron distribution function is given by (Drabble and Goldsmid, 1961)

$$f = f_0 + T\frac{\partial f_0}{\partial E}\tau\,\mathbf{V}.[\mathbf{\nabla}(E_F/T) - E\,\mathbf{\nabla}(1/T)] \tag{3.43}$$

3.10 Electron scattering mechanisms

When electrons flow under the influence of an external electric field or a temperature gradient, they may be scattered by other electrons, phonons or by various impurities and imperfections. Each of these scattering processes can be associated with a resistivity. According to Mathiessen's rule the total resistivity can be obtained simply by adding the separate resistivities. The resistance due to electron-phonon interaction is always present and is one of the most important mechanisms in limiting the carrier mean-free-path. This topic has been dealt with in detail (see, for example, Ziman, 1960) and only a brief description of the basic concepts will be given here.

Electron-phonon scattering is discussed by considering a situation in which an electron in a state $\mathbf{k}_1$ is scattered to a state $\mathbf{k}_2$ as a result of an interaction with a phonon of wave vector $\mathbf{q}$. The problem is to obtain the matrix element between the initial and the final states and calculate the transition probability. In order to simplify the problem, the adiabatic approximation is considered in which the electron wave functions are assumed to keep pace with the ionic motion and the Fermi surface is assumed spherical. The energy conservation condition has to be satisfied and the wave vectors must satisfy

$$\mathbf{k}_1 \pm \mathbf{q} = \mathbf{k}_2 + \mathbf{G} \tag{3.44}$$

The positive and the negative signs correspond to the absorption and the emission of a phonon of wave $\mathbf{q}$. $\mathbf{G}$ refers to the reciprocal lattice vector; $\mathbf{G} = 0$ corresponds to electron-phonon normal processes and $\mathbf{G} \neq 0$ corresponds to umklapp or U-processes.

There is an inherent difficulty in solving the electron Boltzmann equation when phonons act as scattering agents. The scattering term requires a knowledge of the phonon distribution which, in turn, requires a knowledge of the electron distribution. However, in many instances the problem can be simplified if one replaces the phonon distribution by its equilibrium value. The assumption implies that the electron and phonon systems produce entropy independently of each other since they are associated with the departures of corresponding distribution functions from equilibrium. This in turn amounts to neglecting the phonon-drag effects which are a direct consequence of coupling the entropy production of the two systems via the electron-phonon interaction. Further information relating to the effect of phonon-drag on thermal conductivity can be obtained from Ziman (1960) and Seeger (1973). A major topic discussed in this book is the electronic contribution to thermal conductivity which we discuss in terms of the relaxation-time method. Consequently, expressions will be required for the relaxation times for the different scattering processes. Unlike the variational method which gives specific formulae, the relaxation time approach enables general formulae to be obtained for the transport properties. The most widely used technique for obtaining the $\left(\frac{\partial f}{\partial t}\right)_{coll}$ term is the so-called deformation potential method due to Bardeen and Shockley (1950).

In reasonably pure semiconductors (non-degenerate limit) the scattering of electrons by acoustic phonons leads to a relaxation time $\tau = aE^{-1/2}$.

where
$$a = \frac{h^4}{8\pi^3} \frac{\rho v_L^2}{k_B T} \frac{1}{(2m^*)^{3/2} \epsilon_1^2} \qquad (3.45)$$

ρ is the density and v_L is the velocity of the longitudinal phonons. ϵ_1, the deformation potential constant, can be estimated from independent measurements, such as the changes in E_g with pressure and temperature.

In heavily doped semiconductors, scattering by ionized impurities (donor or acceptor) may be important along with phonon scattering, whereas in many polar semiconductors scattering of carriers by polar optical modes is significant.

Generally, the relaxation time for these various scattering mechanisms can be expressed as

$$\tau = aE^s \qquad (3.46)$$

where s takes values of $-\frac{1}{2}, \frac{1}{2}, \frac{3}{2}$ for acoustic scattering, polar optical scattering and ionized-impurity scattering, respectively. The energy independent parameter a takes the value

$$a = \frac{h^2}{2^{1/2} m^{*1/2} e^2 k_B T (\epsilon_\infty^{-1} - \epsilon_0^{-1})} \qquad (3.47)$$

for scattering by polar optical phonons. Here ϵ_∞ and ϵ_0 are the high frequency and static dielectric constants. In polar semiconductors, both acoustic scattering and optical phonon scattering need to be taken into consideration at high temperatures.

In heavily doped semiconductors the scattering of an electron by the Coulomb field of an ionized-impurity atom may become important. The corresponding relaxation time is given by (Smith, 1959)

$$\frac{1}{\tau_I} = \frac{Z^2 e^4 N_I}{16\pi (2m^*)^{1/2} \epsilon^2 E^{3/2}} \ln\left[1 + (2E/E_m)^2\right] \qquad (3.48)$$

Here $-E_m$ is equal to the potential energy of an electron at a distance r_m from an impurity. r_m is the distance at which the Coulomb field ceases to be effective and is approximately equal to half the mean distance between the impurities. N_I is the number of ionized impurities per unit volume and ϵ is the dielectric constant.

Intervalley Scattering

In many-valley semiconductors intervalley scattering of carriers, apart from the usual intravalley scattering, must be taken into consideration. In the transition of an electron from one valley (in k-space) to another a large change in momentum is involved. This momentum may be taken up either by an impurity atom or a phonon near the Brillouin zone boundary where acoustic and optical branches are not far from each other. The case of impurity scattering is limited to low temperatures and very impure semiconductors. As regards scattering by phonons the intra and inter-valley scattering differ in that most of the phonons emitted or absorbed in intravalley scattering have energies considerably lower than the energy of charge carriers, whereas this does not apply to phonons involved in intervalley scattering. Taking into account acoustic-deformation potential scattering in addition to intervalley scattering the total relaxation time is given by

$$\frac{1}{\tau} = \frac{1}{\tau_{ac}} + \frac{1}{\tau_{iv}} \qquad (3.49)$$

where τ_{iv} refers to intervalley scattering. Herring (1955) has calculated the temperature variation of mobility and plotted μ/μ_0 versus T/θ_i, where

$$\mu_0 = \mu_{ac} (T/\theta_i)^{3/2} \qquad (3.50)$$

for different values of W_2/W_1, where W_2 and W_1 refer to the strengths of coupling (of carriers) to intervalley and intravalley modes. Further $\theta_i = \hbar\omega_i/k_B$ and ω_i is the frequency of the intervalley phonon.

In a large number of theoretical calculations the parabolic nature of the energy-momentum relationship described by $E = \hbar^2 k^2/2m^*$, gives a reasonably good description of the electronic behaviour. This is, however, not applicable to small-gap semiconductors and may lead to a serious error in the electronic transport coefficients including electronic thermal conductivity. The carrier effective mass is usually low for small-gap materials and this leads to a small density-of-states. Consequently the band is filled upto relatively higher values of the carrier energy even for moderate concentrations of carriers. At these energies ($E \simeq E_g$) the proximity of the valence band is strongly felt and the non-parabolic nature of the energy-momentum relationship comes into play. The effect of nonparabolicity on the electronic thermal conductivity will be taken up in Chap. IV (Ravich *et al*, 1970, 1971, Zawadzki, 1974, and Bhandari and Rowe, 1984).

Appendix A

Variational approach

The following is a brief description of the variational principle. For further details, the reader is referred to Ziman (1960).

The Boltzmann equation is a linear integral equation with a positive kernel. In its general form the essential features of this equation are described by

$$R(k) = \int \{\phi(k) - \phi(k')\} P(k, k') \, dk' \qquad (A.1)$$

The kernel $P(k, k')$ being a probability, is positive and remains unaltered by the interchange of k and k' (principle of microscopic reversibility). In a more abstract form Eq. A.1 can be written as

$$R = P\phi \qquad (A.2)$$

The operator P transforms the function ϕ into another function R by integration. The inner product of two functions ϕ and ψ is defined by

$$\langle \phi, \psi \rangle = \int \phi(k) \, \psi(k) \, dk \qquad (A.3)$$

Equation A.2, after forming inner products on both sides, is written as

$$\langle \psi, R \rangle = \langle \psi, P\phi \rangle$$

or
$$\langle \phi, R \rangle = \langle \phi, P\phi \rangle \qquad (A.4)$$

The variational principle states that of all functions which satisfy this condition (Eq. A.4), the solution of the integral equation gives $\langle \phi, P\phi \rangle$ its maximum value.

In yet another form the principle can be stated thus: the solution of the integral equation gives to $\langle \phi, P\phi \rangle / \{\langle \phi, R \rangle\}^2$ its minimum value.

The application of this principle to the solution of the Boltzmann equation is straightforward. A trial function comprising known functions and certain arbitrary parameters is defined and the variational function defined by $\langle \phi, P\phi \rangle / \{\langle \phi, R \rangle\}^2$ is calculated for a certain value(s) of the parameter(s). The parameters are then varied until the variational function acquires a minimum value. The trial function with this set of parameters is then the nearest approximation to the true solution.

An interpretation of the variational principle can be derived from steady-state thermodynamics and finds useful application in transport theory. The flow of charge or heat through a material is accompanied by various resistive (scattering) processes, and in the steady-state a regular steady flow of charge or heat is maintained. The resistive (scattering) processes increase the entropy of the system (at a rate S_{scatt}) whereas the field or temperature gradient reduces the local entropy and a balance is maintained. However, entropy once produced cannot be destroyed and appears as macroscopic heat, or as the dissipation of existing temperature gradients, i.e.

$$\dot{S}_{\text{scatt}} = -\dot{S}_{\text{field}} = \dot{S}_{\text{macro}}$$

Appendix B

Thermodynamic concepts*

Having described the generalized forces, flows and various transport coefficients in Sec. 3.8, one may define the rate of entropy production as

$$\dot{S} = \sum_i F_i X_i \tag{B.1}$$

Consider a system of electrons isolated from the surroundings. The thermodynamic states of this system are described by the electrochemical potential† $\bar{\mu}_1$ and the temperature T_1. For another isolated system of electrons $\bar{\mu}_2$ and T_2 may describe the thermodynamic state. When the two systems are brought in thermal as well as diffusive contact, a flow of electrons and energy will begin from one to the other. The total entropy of the combined system will increase with time $(dS/dt > 0)$ nutil an equilibrium is reached.

A somewhat different situation of interest in transport phenomena refers to the steady-state condition. The two systems in thermal and diffusive contact are maintained at their respective $\bar{\mu}$ and T values. In order to maintain a steady state, external reservoirs of heat and electrons are required.

For system 1, the rate of energy change is given as

$$\frac{dE_1}{dt} = 0 = \left(\frac{dE_1}{dt}\right)_{int} + \left(\frac{dE_1}{dt}\right)_{ext} \tag{B.2}$$

$(dE_1/dt)_{int}$ refers to the change due to a flow of energy between the two systems and $(dE_1/dt)_{ext}$ refers to the change as a result of exchange with the external reservoir. A similar equation can be written for E_2, N_1 and N_2. The conservation of energy and electrons requires

$$(dE_1/dt)_{int} + (dE_2/dt)_{int} = 0$$
$$(dN_1/dt)_{int} + (dN_2/dt)_{int} = 0 \tag{B.3}$$

The rate of entropy production due to the internal processes can be written as

$$(dS/dt)_{int} = \frac{1}{T_1}(dE_1/dt)_{int} - \frac{\bar{\mu}_1}{T_1}(dN_1/dt)_{int} + \frac{1}{T_2}\left(\frac{dE_2}{dt}\right)_{int} - \frac{\bar{\mu}_2}{T_2}\left(\frac{dN_2}{dt}\right)_{int}$$

With the use of Eq. B.3, one obtains

$$\left(\frac{dS}{dt}\right)_{int} = (dE_1/dt)_{int}\left(\frac{1}{T_1} - \frac{1}{T_2}\right) - (dN_1/dt)_{int}\left(\frac{\bar{\mu}_1}{T_1} - \frac{\bar{\mu}_2}{T_2}\right) \tag{B.4}$$

The reservoir connected to system 1 supplies energy at a rate

$$(dE_1/dt)_{ext} = -(dE_1/dt)_{int}$$

and electrons at a rate $(dN_1/dt)_{ext} = -(dN_1/dt)_{int}$. The entropy supplied by the reservoir is

$$\left(\frac{dS_1}{dt}\right)_{ext} = \frac{1}{T_1}(-dE_1/dt)_{int} - \frac{\bar{\mu}_1}{T_1}(-dN_1/dt)_{int}$$

---

*For details see Callen (1948) and Drabble and Goldsmid (1961).

†Also referred to as the Fermi energy in Sec. 3.9.

Similarly, for system 2 in contact with another reservoir

$$\left(\frac{dS_2}{dt}\right)_{ext} = \frac{1}{T_2}(-dE_2/dt)_{int} - \frac{\bar{\mu}_2}{T_2}(-dN_2/dt)_{int}$$

$$= \frac{1}{T_2}(dE_1/dt)_{int} - \frac{\bar{\mu}_2}{T_2}(dN_1/dt)_{int}$$

Finally $\quad (dS/dt)_{ext} = \left(\frac{1}{T_2} - \frac{1}{T_1}\right)(dE_1/dt)_{int} - \left(\frac{\bar{\mu}_2}{T_2} - \frac{\bar{\mu}_1}{T_1}\right)(dN_1/dt)_{int}$

(B.5)

This gives

$$(dS/dt)_{ext} + (dS/dt)_{int} = 0$$

The combined system 1−2 produces entropy at a certain rate as a result of internal processes, such as scattering, which is removed by the external sources.

This result can be generalized for the situation where $\bar{\mu}$ and T vary continuously. In a small volume element the entropy production per second per unit volume is balanced by the rate at which it is removed from it. The rate of internal entropy production is then given by

$$(dS/dt)_{int} = \mathbf{W} \cdot \nabla(1/T) - \mathbf{J} \cdot \nabla(\bar{\mu}/T) \quad \text{(B.6)}$$

W and **J** refers to the flow density vectors for energy and electrons.

The choice of "forces" having been made the "flows" can be expressed in terms of the "forces" as in Eq. 3.24.

References

Bardeen, J. and Shockley, W. (1950), *Phys. Rev. 80*, 72.
Bhandari, C.M. and Rowe, D.M. (1984), *9th European Thermophysical Proper. Conf*, Manchester, September 1984, also in *High Temperatures—High Pressures* (1985) *17*, 97.
Blatt, F.J, (1957), *Solid State Physics 4*, 199 (ed. F. Seitz and D. Turnbull), Academic Press, New York
Bube, R.H. (1974), *Electronic Properties of Crystalline Solids*, Academic Press, New York.
Busch, G. and Schade, H. (1976), *Lectures on Solid State Physics*, Pergamon Press, London.
Callen, H.B. (1948), *Phys. Rev. 73*, 1349.
Drabble, J.R. and Goldsmid. H.J. (1961), *Thermal Conduction in Semiconductors*, Pergamon Press, London.
Drude, P. (1900), *Annln. Phys. 1*, 566.
Drude, P. (1902), *Annln. Phys. 7*, 687.
Herring, C. (1955), *Bell Syst. Tech. J. 34*, 237.
Fistul, V.I. (1969), *Heavily Doped Semiconductors*, Plenum Press, New York p. 154.
Hutner, R.A., Rittner, E.S. and DuPre, F.K. (1950), *Philips Res. Rep. 5*, 188.

Kittel, C. (1976), *Introduction to Solid State Physics* (5th ed.), John Wiley, New York.

Ravich, Yu. I., Efimova, B.A. and Smirnov, I.A. (1970), *Semiconducting Lead Chalcogenides* (ed. L.S. Stil, bans) Plenum Press, New York.

Ravich, Yu. I., Efimova, BA. and Tamarchenko, V.I. (1971) *Phys. Stat. Solidi (b)* 43, 11.

Seeger, K. (1973), Semiconductor Physics, *Springer Verlag*, New York.

Smith, R.A. (1959), Semiconductors, *Cambridge University Press*.

Wilson, A.H. (1953), *The Theory of Metals* (2nd ed.), Cambridge University Press.

Zawadzki, W. (1974), *Adv. in Phys.* 23, 437.

Ziman, J.M. (1960), *Electrons and Phonons*, Clarendon Press, Oxford.

Chapter IV

Electronic Thermal Conductivity

4.1 Introduction

The basic features of the electronic behaviour of solids, and in particular of semiconductors, have been outlined in Chap. III. This chapter deals with the various aspects of the electronic contribution* to the thermal conductivity of semiconducting materials. The discussion essentially refers to crystalline semiconductors although it can be extended to include amorphous and liquid semiconductors (see Chapter IX).

The electronic contribution to the thermal conductivity λ_e is insignificant at very low doping levels but becomes increasingly important as the doping level increases; for example it is around 25 per cent of the total thermal conductivity in silicon-germanium alloys with a carrier concentration of 10^{25} m^{-3}. It is customary and also convenient to express the various electronic transport parameters as a function of the reduced Fermi energy ξ. Calculation of λ_e requires evaluation of the Lorenz factor $\mathcal{L}$ (or the Lorenz number $L = (k_B/e)^2 \mathcal{L}$) defined by $\lambda_e = (k_B/e)^2 \mathcal{L} \sigma T$ and Sec. 4.2-4.5 are devoted to the problem of obtaining $\mathcal{L}$ in the extrinsic and intrinsic semiconduction ranges and for various carrier scattering mechanisms.

The usual semiconductor model based upon parabolic energy bands gives a reasonably good agreement between theory and experimental data except for materials with narrow energy band gaps—a situation which was encountered in the analysis of the transport properties of lead chalcogenides (Ravich *et al*, 1970, 1971). Section 4.6 is devoted to the effect of non-parabolic energy bands on the Lorenz factor.

Generally magnetic fields affect thermal conductivity through the electronic contribution (Sec. 4.7) although in some special cases it can be influenced in other ways as discussed in Chap. VIII and X.

4.2 Extrinsic semiconductors—evaluation of various electronic transport coefficients using the relation time approximation

---

*The contribution to the thermal conductivity due to heat being transported by electrons in *n*-type semiconductors or holes in *p*-type material is generally referred to as electronic contribution.

Most semiconductors employed in device applications are extrinsic with suitable amounts of dopant, either p or n-type, having been introduced in order to optimise the material performance. In any theoretical analysis of thermal conductivity the electronic contribution must be separated from the lattice component; the latter can then be analysed in terms of phonon transport.

The theoretical calculation of the electronic contribution can readily be carried out within the framework of the relaxation time approximation (see for example, Drabble and Goldsmid, 1961).

As described in Sec. 3.10, the carrier relaxation time τ is given by $\tau = aE^s$ where a depends upon temperature and the scattering parameter s takes values of $-\frac{1}{2}$, $\frac{1}{2}$ and $\frac{3}{2}$ for scattering by acoustic phonons, polar optical phonons and ionized-impurity scattering, respectively.

The components of the flow of electrons and energy (of the electron system) are given by

$$-J_i = -\frac{1}{4\pi^3}\int V_i f\, d\mathbf{k}$$

and

$$(W_e)_i = \frac{1}{4\pi^3}\int E V_i f\, d\mathbf{k} \tag{4.1}$$

The distribution function f for electrons obtained by solving the Boltzmann equation can be substituted in these equations which can then be written in a form similar to that of Eq. (3.24). The coefficients $L_{ik}^{(m)}$ can be expressed in terms of integrals of the type (Drabble and Goldsmid, 1961)

$$I_{ij}^{(l)} = \frac{T}{4\pi^3}\int E^l\, \tau V_i V_j \frac{\partial f_0}{\partial E}\, d\mathbf{k} \tag{4.2}$$

$$L_{ij}^{(1)} = -I_{ij}^{(0)},\ L_{ij}^{(2)} = L_{ij}^{(3)} = I_{ij}^{(1)}$$

and

$$L_{ij}^{(4)} = -I_{ij}^{(2)} \tag{4.3}$$

If a single spherical band with an isotropic effective mass is considered, the calculation of the transport coefficients is considerably simplified. The carrier energy E and the relaxation-time can then be taken to depend on $|\mathbf{k}|$.

$$d\mathbf{k} = k^2\, dk\, \sin\theta\, d\theta\, d\phi$$

the integral over angular coordinates is zero unless $i = j$, in which case it becomes equal to $4\pi/3$. Under these conditions appropriate to an isotropic solid, $L^{(n)}$ are reduced to scalars. The integral over the modulus k of $\mathbf{k}$ can be transformed to an integral with the variable $E = \hbar^2 k^2/2m^*$, the energy being measured with respect to the band edge.

Introducing a minus sign for simplicity in subsequent formulation these integrals can be related to the following integrals (Drabble and Goldsmid, 1961).

$$M_l = -\frac{1}{m^*}\left(\frac{2m^*}{\hbar^2}\right)^{3/2}\frac{T}{3\pi^2}a\int E^l E^{(s+3/2)}\frac{\partial f_0}{\partial E}\, dE \tag{4.4}$$

An integration by parts further simplifies M_l to

$$M_l = \frac{a}{m^*} \frac{T}{3\pi^2} \left(\frac{2m^* k_B T}{\hbar^2}\right)^{3/2} \left(l + s + \frac{3}{2}\right) (k_B T)^{(l+s)} F_{s+l+1/2}(\xi) \quad (4.5)$$

where F are the Fermi integrals,

$$F_r(\xi) = \int_0^\infty \frac{\eta^r \, d\eta}{1 + \exp[(\eta - \xi)]}$$

$$\eta = \frac{E}{k_B T} \quad \text{and} \quad \xi = \frac{E_F}{k_B T}$$

The coefficinets $L_{ik}^{(m)}$ can now be expressed in terms of integrals M_l and related to the various transport parameters, such as the electrical conductivity σ, the Peltier coefficient π, the Seebeck coefficient α and the electronic thermal conductivity λ_e (for details see Drabble and Goldsmid, 1961). Of interest here is the electronic thermal conductivity given by

$$\lambda_e = \frac{1}{T^2} L_{11}^{(4)} - \frac{1}{T} \sigma \left(\pi - \frac{\bar{\mu}}{e}\right)^2 \quad (4.6)$$

where $\bar{\mu}$ is the electrochemical potential defined in Sec. 3.8 of the previous chapter. In terms of the integrals M_l one obtains

$$\lambda_e = \frac{M_0}{T^2} \left\{ \frac{M_2 M_0 - M_1^2}{M_0^2} \right\} \quad (4.7)$$

4.3 Evaluation of the Lorenz factor $\mathcal{L}$

The basic problem in the theoretical determination of λ_e is in practice reduced to the evaluation of $\mathcal{L}$. In a strictly theoretical calculation of λ_e it is the product $\mathcal{L}\sigma'$ rather then $\mathcal{L}$ which needs to be calculated. Defining a reduced electrical conductivity by $\sigma' = \left(\frac{k_B}{e}\right)^2 \frac{T}{\lambda_L} \sigma$, it is of interest to obtain $\mathcal{L}\sigma' = \left[\frac{\lambda_e}{\lambda_L}\right]$ as a function of ξ for various scattering mechanisms and these calculations form the theme of subsection 4.6.5.

The Lorenz factor for a parabolic energy band and a particular scattering mechanism described by a parameter s is given by

$$\mathcal{L} = \left\{ \frac{(s + \frac{7}{2}) F_{s+5/2}(\xi)}{(s + \frac{3}{2}) F_{s+1/2}(\xi)} - \delta^2(\xi) \right\} \quad (4.8)$$

where

$$\delta(\xi) = \frac{(s + \frac{5}{2}) F_{s+3/2}(\xi)}{(s + \frac{3}{2}) F_{s+1/2}(\xi)} \quad (4.9)$$

In the non-degenerate limit $-\xi \gg 1$

$$F_r(\xi) = \exp(\xi) \Gamma(r + 1)$$

and this leads to the result

$$\mathcal{L} = (s + \tfrac{5}{2}) \quad (4.10)$$

In the degenerate limit $\xi \gg 1$, the Fermi integrals can be expressed as a series which converge rapidly. Retaining only the first and the significant term gives:

$$\mathcal{L} = \frac{\pi^2}{3} \qquad (4.11)$$

In general for arbitrary ξ values $\mathcal{L}$ as a function of ξ can be obtained by a numerical evaluation of Eq. 4.8 and the results of calculation for $s = -\frac{1}{2}, +\frac{1}{2}$ and $+\frac{3}{2}$ are given in Fig. 4.1. It is evident that $\mathcal{L}$ values for different scattering mechanisms differ considerably at low carrier concentrations and converge to the limiting value $\pi^2/3$ for $\xi \gg 1$.

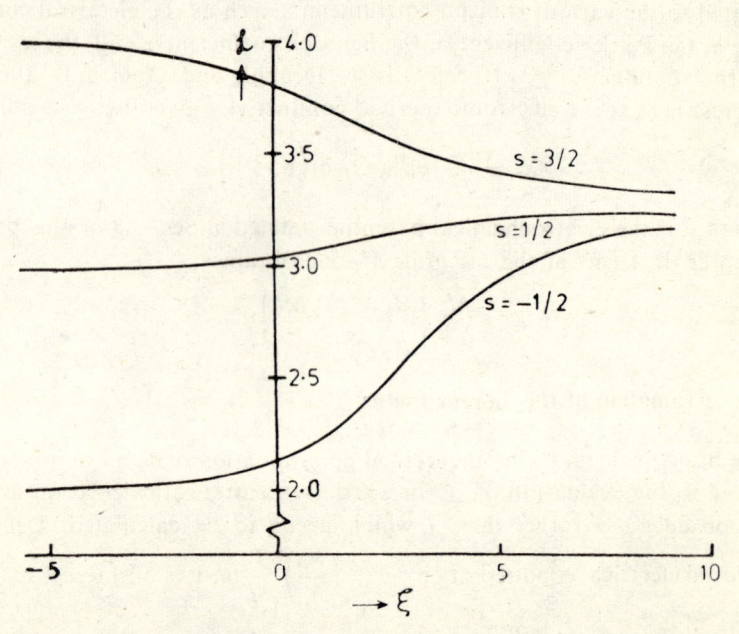

Fig. 4.1 Lorenz factor as a function of reduced Fermi energy for a parabolic band for $s = -1/2, +1/2$ and $3/2$.

4.4 Mixed scattering—acoustic scattering and ionized-impurity scattering operative simultaneously

The Lorenz factor has been obtained in the previous section for various individual scattering mechanisms. In heavily doped semiconductors several scattering mechanisms may operate simultaneously; in a number of covalent semiconductors acoustic phonon scattering accompanies ionized-impurity scattering and in polar semiconductors optical phonon scattering must also be considered.

In the case of mixed scattering, when acoustic scattering and ionized-impurity scattering act simultaneously, the associated relaxation times τ_{ac} and τ_{imp} are given by

$$\tau_{ac} = \tau_{oL}\, \eta^{-1/2} \tag{4.12}$$

$$\tau_{imp} = \tau_{oi}\, \eta^{3/2}$$

Here η is the reduced carrier energy $E/k_B T$. The combined relaxation time τ is obtained by adding the inverse relaxation times for the two mechanisms. Thus

$$\tau = \tau_{oL}\, \eta^{3/2} / (\eta^2 + b^2) \tag{4.13}$$

where $b^2 = \tau_{oL}/\tau_{oi}$ is a measure of the relative strengths of ionized-impurity scattering and acoustic scattering. The Lorenz factor in this case is given by (Fistul, 1969)

$$\mathcal{L} = (\phi_3 \phi_5 - \phi_4^2)/\phi_3^2 \tag{4.14}$$

where

$$\phi_n(\xi, b) = \int_0^\infty \frac{\eta^n \exp(\eta - \xi)\, d\eta}{(\eta^2 + b^2)\{1 + \exp(\eta - \xi)\}^2} \tag{4.15}$$

where n is an integer.

The limiting cases can readily be obtained from these expressions

(a) *Acoustic phonon scattering dominant* $(b^2 \ll \bar{\eta}^2)$

$$\mathcal{L}_{ac} = \frac{3 F_0(\xi) F_2(\xi) - 4 F_1^2(\xi)}{F_0^2(\xi)} \tag{4.16}$$

In the lightly doped material $(\xi \ll 0)$

$$\mathcal{L}_{ac} = 2 \tag{4.17}$$

and for heavily doped material $(\xi \gg 0)$

$$\mathcal{L}_{ac} = \frac{\pi^2}{3}\left(1 - \frac{\pi^2}{3\xi^2}\right) \tag{4.18}$$

(b) *Ionized-impurity scattering dominant* $(b^2 \gg \bar{\eta}^2)$

$$\mathcal{L}_{imp} = \frac{15 F_2(\xi) F_4(\xi) - 16 F_3^2(\xi)}{9 F_2^2(\xi)} \tag{4.19}$$

In lightly doped material $(\xi \ll 0)$

$$\mathcal{L}_{imp} = 4 \tag{4.20}$$

and for heavily doped material $(\xi \gg 0)$

$$\mathcal{L}_{imp} = \frac{1}{\xi^2 + \pi^2}\left[\xi^2 + \frac{10\pi^2}{3} - \frac{(\xi^3 + 2\pi^2)^2}{\xi^2 + \pi^2}\right] \tag{4.21}$$

Fig. 4.2 is a plot of $\mathcal{L}$ versus ξ for various values of the parameter b; $b = 0$ corresponds to acoustic scattering only and as b increases the ionized-impurity scattering becomes more significant. Appropriate values of b and ξ for a particular material can be determined from measurements of Hall mobility and Seebeck coefficient (Rowe and Bhandari, 1983).

4.5 Intrinsic semiconduction

The expressions obtained in the last section correspond to a model in which only charge carriers in one band contribute to the transport properties, the

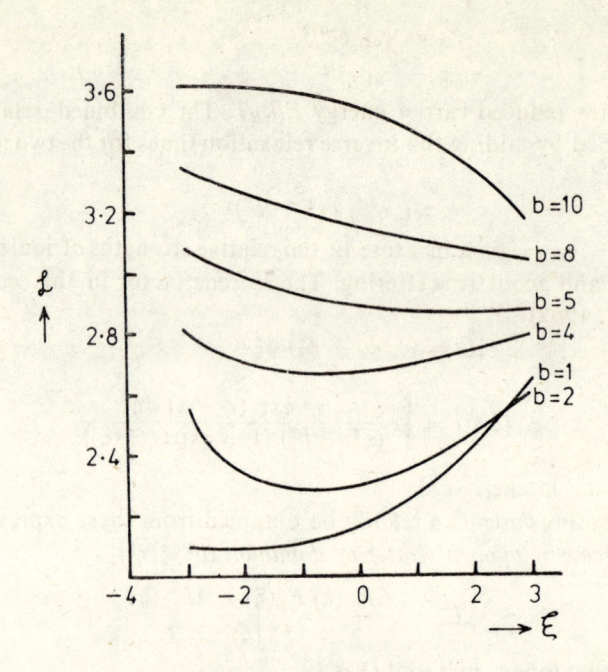

Fig. 4.2 Lorenz factor $\mathscr{L}$ as a function of ξ and b for acoustic phonon scattering and ionized-impurity scattering acting simultaneously.

Fermi level being much closer to one band than to the other. However, at sufficiently high temperatures the Fermi level is near the middle of the energy gap and carriers in both bands contribute to the transport properties. When electrons and holes are present in almost equal concentrations they flow down a temperature gradient carrying all their energy and yet give rise to a vanishingly small electric current. Each electron-hole pair can effectively transport a quantity of energy equal to the energy gap and this would give rise to a very large value of L (see Parrott, 1979). Electron-hole pairs are created at the hot end by the absorption of energy from the source. These pairs then move down the temperature gradient and recombine releasing the recombination energy. This process is sometimes called bipolar thermodiffusion (Price 1955).

If we write the Lorenz number for electrons and holes as L_e and L_h where $L_e \sigma_e T = \lambda_e$ and $L_h \sigma_h T = \lambda_h$, then the total Lorenz number, taking into account the bipolar thermodiffusion, is given by (Drabble and Goldsmid, 1961):

$$L = \frac{L_e \sigma_e}{\sigma_t} + \frac{L_h \sigma_h}{\sigma_t} + \left(\frac{k_B}{e}\right)^2 \frac{\sigma_e \sigma_h}{(\sigma_e + \sigma_h)^2} (\delta_n + \delta_p + \xi_g)^2 \quad (4.22)$$

where $\delta_{n,p}$ ($\xi_{n,p}$) are as given by Eq. (4.9), and $\sigma_t = \sigma_e + \sigma_h$. L_e and L_h are of the order of $2(k_B/e)^2$ and $\delta_n = \delta_p = s + 5/2$, in the non-degenerate case, $-\xi \gg 1$.

For acoustic scattering ($s = -1/2$) one obtains

$$L = 2\left(\frac{k_B}{e}\right)^2 + \left(\frac{k_B}{e}\right)^2 \frac{\sigma_e \sigma_h}{(\sigma_e + \sigma_h)^2}(4 + \xi_g)^2 \qquad (4.23)$$

The bipolar term is usually negligible if either $\sigma_e \gg \sigma_h$ or $\sigma_e \ll \sigma_h$ but when $\sigma_e = \sigma_h$, it becomes very large. At around 1200 K the energy gap of silicon is $\sim 10\,k_B T$ and in the intrinsic range when σ_e and σ_h are almost equal one obtains

$$L \simeq 50\left(\frac{k_B}{e}\right)^2 \qquad (4.24)$$

which is very large compared to the usual value of about $2(k_B/e)^2$. This effect can be seen in the experiments of Fulkerson *et al* (1968) on the thermal conductivity of Si at high temperatures (Fig. 4.3). In bismuth telluride, $\xi_g = 5$ at 300 K and for $s = -1/2$, L has a value of about $22(k_B/e)^2$ in the intrinsic range; a result in good agreement with experimental data.

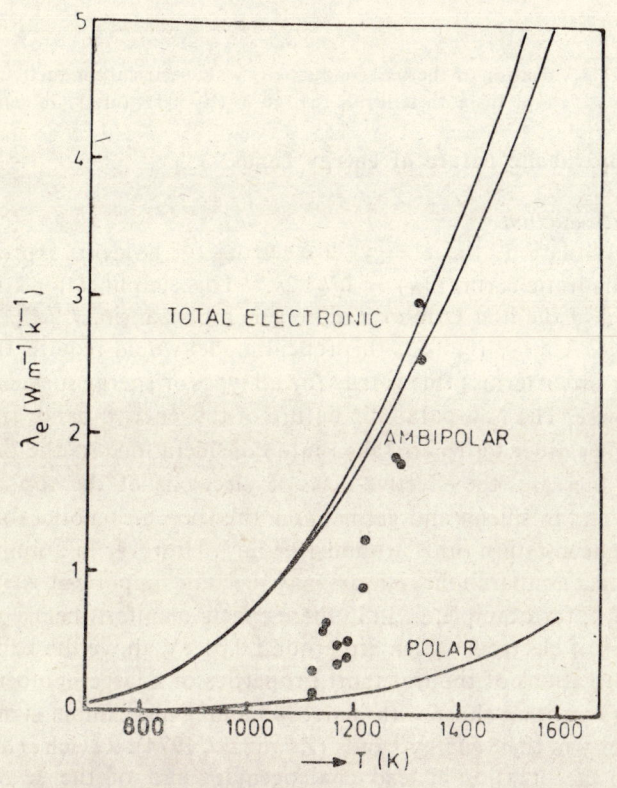

Fig. 4.3 Thermal conductivity versus temperature for silicon in the intrinsic range (after Fulkerson *et al* 1968).

If the total thermal conductivity λ is plotted against the electrical conductivity σ for different carrier concentrations λ_L can be determined by extrapolating the curves to $\sigma = 0$. Goldsmid (1958 and 1965) adopted this procedure for Bi$_2$Te$_3$ at 150 K. At 300 K the bipolar term becomes important and λ rapidly rises in the intrinsic region (Fig. 4.4).

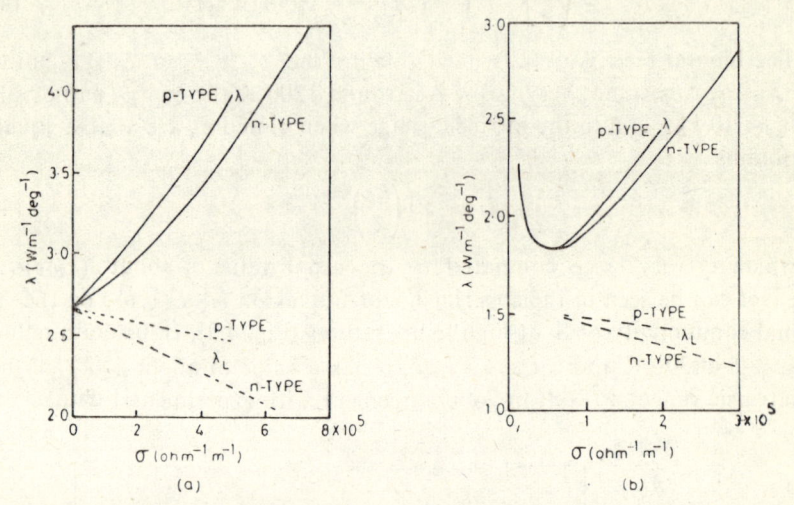

Fig. 4.4 Variation of thermal conductivity with electrical conductivity in p-and n-type bismuth telluride (a) 150 K (b) 300 K (after Goldsmid, 1958).

4.6 Nonparabolic nature of energy bands

4.6.1 *Introduction*

In many situations the energy of electrons (or holes) is expressible in the simple quadratic form, $E(k) = \hbar^2 k^2/2m^*$. This simplification arises from the utilization of the first term of a more general expansion of $E(k)$ about the band-edge. A more rigorous theoretical model would require the inclusion of higher order terms; this is true for all types of energy surfaces—spherical or otherwise. The non-parabolic nature of the energy band appears when these higher order terms are taken into consideration. As the population of the band increases the effective-mass of electrons at the top of the band changes, and in silicon and germanium this becomes noticeable when the carrier concentration (n) is around 10^{26} m^{-3}. However in a number of semiconductors, nonparabolic effects may become important at much lower values of n; for example, in InSb these effects manifest themselves upon the behaviour of electrons which are around 0.01 eV above the band edge.

Investigations of the transport properties of a large number of semiconductors reveal that those with narrow band gaps exhibit significant nonparabolicity in their energy bands (Zawadzki, 1974). Ravich *et al* (1971) have discussed the situation in lead chalcogenides and of the several methods available for use in investigating the nonparabolicity of the conduction and valence bands in these materials they emphasize the analysis of thermoelectric power (Seebeck effect) in a strong magnetic field. This method gives information about density-of-states versus energy and chemical-potential versus carrier concentration. A comparison of the results obtained by this technique with the theoretical predictions leads to the conclusion that a two-band model is appropriate for the chalcogenides.

In obtaining expressions for the transport coefficients Ravich *et al*, adopted an approach similar to the theory developed by Kane (1957) for InSb in which the electron energy dispersion is taken into account by the expression (Ravich *et al*, 1970)

$$\frac{\hbar^2}{2}\frac{k_T^2}{m_{TO}^*} + \frac{\hbar^2}{2}\frac{k_L^2}{m_{LO}^*} = E\left(1 + \frac{E}{E_g}\right) \quad (4.25)$$

k_L and k_T are the longitudinal and transverse components of the wave vector (electron or hole) and m_{LO}^* and m_{TO}^* are the longitudinal and transverse components of the effective-mass tensor near the band extremum. The effective-mass

$$\frac{1}{m_j^*} = \frac{1}{\hbar^2}\frac{1}{k_j}\frac{\partial E}{\partial k_j} \quad (4.26)$$

is now energy-dependent with

$$m_j^* = m_{j0}^*\left(1 + 2\frac{E}{E_g}\right) \quad (4.27)$$

4.6.2 *The effect of nonparabolic energy bands on the Lorenz number*

Nonparabolicity of electron energy bands has an effect on the various transport coefficients and a number of useful formulae are reported in the literature (Smirnov and Ravich, 1967, and Ravich *et al*, 1970). Of relevance to the theme of this book is the effect of nonparabolicity on the electronic contribution to thermal conductivity. The evaluation of the Lorenz-factor as a function of the reduced Fermi-potential ξ for different carrier scattering mechanisms has been discussed earlier for parabolic bands. For nonparabolic energy bands the Lorenz factor can be expressed as

$$\mathcal{L} = \frac{\int A(\eta)\, B(\eta)\eta^2\, d\eta}{\int A(\eta)\, B(\eta)\, d\eta} - \delta^2 \quad (4.28)$$

where $A(\eta) = \left\{-\frac{\partial f}{\partial \eta}\tau(\eta)\right\}$; $B(\eta) = \{\eta(1+\beta\eta)\}^{3/2}/(1+2\beta\eta)$

and

$$\delta(\xi) = \frac{\int A(\eta)\, B(\eta)\, \eta\, d\eta}{\int A(\eta)\, B(\eta)\, d\eta} \quad (4.29)$$

$\beta = k_B T/E_g$ is the inverse of the reduced energy gap; f is the Fermi function.

4.6.3 *Scattering by acoustic phonons*

In order to evaluate the expressions which occur in the Lorenz factor the quantity $\tau(\eta)$ must be known. It can be written as (Smirnov and Ravich, 1967)

$$\tau \sim \frac{1}{|M|^2 \rho(\eta)} \qquad (4.30)$$

where M is the matrix element for the electron-phonon interaction and $\rho(\eta)$ is the density of-states. The expression is simplified if it is assumed that the matrix element M depends upon energy in the same way as for a parabolic band. For acoustic scattering this amounts to $|M|^2 =$ constant. The relaxation time is then given by

$$\tau_{ac} \sim \frac{1}{\rho(\eta)} \propto \frac{1}{\{\eta(1+\beta\eta)\}^{1/2}(1+2\beta\eta)} \qquad (4.31)$$

An expression for $\tau(\eta)$ which includes the energy dependence of M has been discussed by Ravich *et al* (1971).

Sometimes it is convenient to write $\mathcal{L}$ in terms of the generalized Fermi integrals ${}^n L_l^m (\xi, \beta)$

$$\mathcal{L} = \frac{{}^2L_{-2}^1 \, {}^0L_{-2}^1 - \left({}^1L_{-2}^1\right)^2}{\left({}^0L_{-2}^1\right)^2} \qquad (4.32)$$

where
$$\,^n L_l^m (\xi, \beta) = \int_0^\infty \left(-\frac{\partial f}{\partial \eta}\right) \eta^n \, [\eta(1+\beta\eta)]^m \, (1+2\beta\eta)^l \, d\eta \qquad (4.33)$$

For acoustic scattering a plot of $\mathcal{L}$ versus ξ is given in Fig. 4.5 for different values of the parameter β.

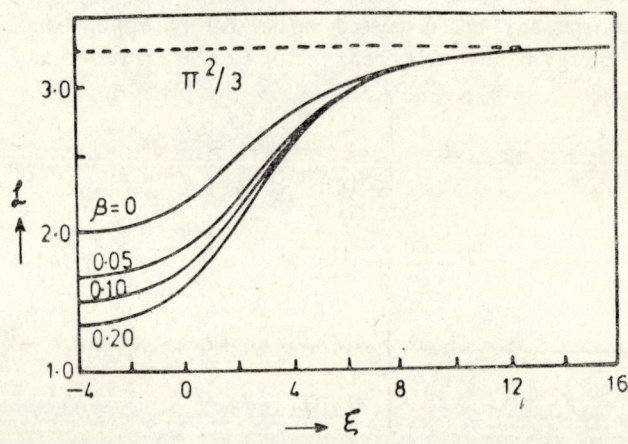

Fig. 4.5 Lorenz factor $\mathcal{L}$ as a function of ξ for various β values considering only acoustic-phonon scattering (after Smirnov and Ravich, 1967).

4.6.4 *Polar-optical scattering*

Scattering of charge carriers by polar-optical modes must be taken into consideration in an analysis of carrier scattering in polar semiconductors. The carrier relaxation time for a simple parabolic band is readily obtained and is given by

$$\tau_{op}^{-1} = \frac{2k_B T e^2 (\epsilon_\infty^{-1} - \epsilon_0^{-1})}{\hbar^2 V} \qquad (4.34)$$

where V is the carrier velocity, e is its charge and ϵ_∞ and ϵ_0 are the high frequency and static dielectric constants, respectively.

For a nonparabolic 'Kane band', including the effect of screening of the electric field produced by optical vibrations, the carrier relaxation time is given by (Ravich et al., 1971)

$$\tau_{op}^{-1} = \frac{2^{1/2} e^2 k_B T m_d^{*1/2}}{\hbar^2 E^{1/2}} (\epsilon_\infty^{-1} - \epsilon_0^{-1}) \frac{1 + 2(E/E_g)}{[1 + (E/E_g)]^{1/2}}$$

$$\times \left\{ \left[1 - \delta_\infty \ln\left(1 + \frac{1}{\delta_\infty}\right) \right] - \frac{2E(E_g + E)}{(E_g + 2E)^2} \right.$$

$$\left. \times \left[1 - 2\delta_\infty + 2\delta_\infty^2 - \ln\left(1 + \frac{1}{\delta_\infty}\right) \right] \right\} \qquad (4.35)$$

δ_∞ stands for $(2kr_\infty)^{-2}$ and r_∞, the screening radius for a medium with dielectric constant ϵ_∞, is given by

$$r_\infty^{-2} = \frac{2^{5/2} e^2 m_d^{*3/2}}{\pi \hbar^3 \epsilon_\infty} \int_0^\infty \left(-\frac{\partial f}{\partial E}\right) \left(E + \frac{E^2}{E_g}\right)^{1/2} \left(1 + 2\frac{E}{E_g}\right) dE \qquad (4.36)$$

In many situations kr_∞ is very close to unity.

If screening effects are ignored, the Lorenz factor can again be written in terms of the generalized Fermi integrals (Ravich et al, 1970).

$$\mathcal{L} = \left\{ {}^2L_{-2}^2 \middle/ {}^0L_{-2}^2 \right\} - \delta^2 \qquad (4.37)$$

where
$$\delta = {}^1L_{-2}^2 \middle/ {}^0L_{-2}^2 .$$

A similar expression can be written when screening effects are taken into account and the integrals in this case are given by:

$$^nL_l^m = \int_0^\infty \frac{\left(\frac{\partial f}{\partial \eta}\right) \eta^n [\eta(1 + \beta\eta)]^m (1 + 2\beta\eta)^l \, d\eta}{[1 - \delta_\infty \ln(1 + 1/\delta_\infty)] - \frac{2\beta\eta(1 + \beta\eta)}{(1 + 2\beta\eta)^2} [1 - 2\delta_\infty + 2\delta_\infty^2 \ln(1 + 1/\delta_\infty)]}$$

(4.38)

4.6.5 *Electronic contribution to the thermal conductivity*

The equations discussed above can be solved to obtain the dependence of the Lorenz number on ξ for the scattering mechanisms considered. It is also possible to obtain $\mathcal{L}\sigma'$ ($= \lambda_e/\lambda_L$), where σ' is the reduced electrical conductivity defined earlier (see Sec. 4.3). Employing the appropriate expressions

70 THERMAL CONDUCTION IN SEMICONDUCTORS

in the non-parabolic case (Ravich *et al*, 1970) λ_e/λ_L can be evaluated. Figure 4.6 gives such a theoretical plot for $\beta = 0.08$, which corresponds to the realistic value of β for PbTe at room temperature. Corresponding values for the Lorenz factor $\mathcal{L}$ are also shown on a different scale. The observed values of λ_e/λ_L reported in the literature (Wright, 1970) can only be explained if nonparabolicity is included. The effect of nonparabolic energy bands on the electronic thermal conductivity of a number of other narrow-gap

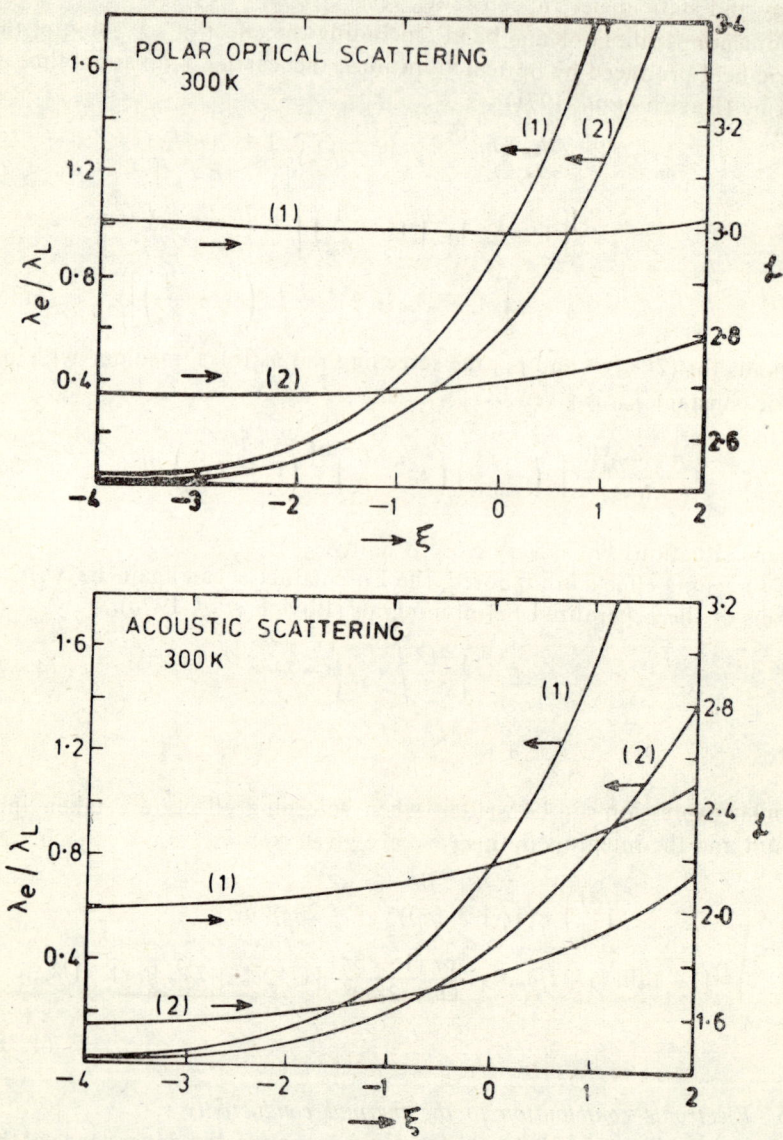

Fig. 4.6 Lorenz factor $\mathcal{L}$ and the ratio λ_e/λ_L as a function of ξ for acoustic scattering and polar optical scattering in PbTe (after Bhandari and Rowe, 1985). Curves [1] and [2] refer to parabolic and nonparabolic bands, respectively.

semiconductors has been reported by Bhandari and Rowe (1984 and 1985).

Various electronic transport coefficients may be modified if the multivalley structure of the semiconducting material is taken into account. The net effect of introducing a multivalley structure may be quite significant in the evaluation of some transport coefficients, such as electrical conductivity and Seebeck coefficient. Evaluation of electrical conductivity, and hence electronic thermal condutivity in the framework of multivalley structure will require the inclusion of intervalley scattering of carriers along with the usual intravalley scattering (Herring, 1955, and Seeger, 1982).

4.7 Thermomagnetic effects

4.7.1 *Introduction*

The effect of a magnetic field on thermal conductivity arises essentially due to its effect on λ_e. The magnetic field can be used to suppress the electronic component and thus enable it to be separated from the lattice thermal conductivity λ_L. The dependence of thermal conductivity on the magnetic field can be expressed as

$$\lambda(B) = \lambda_L + L(B)\,\sigma(B)T \tag{4.39}$$

Except under special circumstances λ_L can be taken to be independent of the magnetic field while L and σ in general depend upon it. If it is assumed that the Lorenz number is independent of the field, a plot of $\lambda(B)$ against $\sigma(B)T$ can be used to obtain λ_L. White and Woods (1958) applied this method to antimony.

In semiconductors the effect of a magnetic field on thermal conductivity is in general small as most of the heat is conducted by lattice waves. However, in degenerate or partially degenerate materials thermomagnetic effects can be observed. The electric and magnetic fields and temperature gradient can be applied in various ways to observe the Nernst, Righi-Leduc, Ettinghausen and Hall effects (Wilson, 1953, and Ziman, 1960).

To observe the various effects the sample under investigation is taken in the same shape as in Hall effect measurements. A magnetic field B_z is applied along the z-direction while a temperature gradient $\partial T/\partial x$ is applied along the x-direction. Just as in Hall effect measurements a transverse field ϵ_y is developed. The intensity of this field (the Nernst field) is given by

$$\epsilon_y = Q_N \left(\frac{\partial T}{\partial x}\right) B_z \tag{4.40}$$

Q_N defines the Nernst coefficient and is measured in units of m²/sK.

In addition a temperature gradient will develop along y-direction and is given by

$$\frac{\partial T}{\partial y} = S_{RL} \frac{\partial T}{\partial x} B_z \tag{4.41}$$

S_{RL} defines the Righi-Leduc coefficient and is measured in m²/Vs.

In the Hall effect, a transverse voltage ϵ_y is measured in the presence of a magnetic field B_z and an electric current j_x. However, in some situations this may be subject to an error as the heat transported by carriers deflected by the magnetic field generates a transverse temperature gradient given by

$$\frac{\partial T}{\partial y} = P_E j_x B_z \tag{4.42}$$

where P_E is the Ettinghausen coefficient.

4.7.2 Separation of electronic and lattice components

It is possible, in principle, to separate the electronic and lattice contributions to thermal conductivity by suppressing the electronic contribution with a suitably oriented magnetic field. However, the required magnetic field must be large enough to ensure that $\mu^2 B^2 \gg 1$ (in SiGe alloys with $\mu \sim .05$ m^2 V^{-1} s^{-1}, $B \sim 50$–60 T).

Such large magnetic fields are difficult to obtain and techniques have been devised to obtain information about high field effects from measurements made at relatively low fields.

Korenblit *et al* (1968) developed a semiempirical theory of thermomagnetic effects and applied it to InSb. Armitage and Goldsmid (1969) analysed the thermomagnetic behaviour of Cd$_3$As$_2$. A thermal mobility μ_T is employed and is defined by

$$\mu_T = (L/L_0') \mu_H$$

where μ_H is the Hall mobility and L the Lorenz number. L_0' refers to the situation when only elastic scattering is present. The thermal conductivity in the presence of a magnetic field is given by

$$\lambda = \lambda_L + \frac{\lambda_e (B=0)}{1 + \mu_T^2 B^2} \tag{4.43}$$

Consequently,

$$-\frac{B^2}{\Delta\lambda} = \frac{1}{\mu_T \lambda_e} + \frac{B^2}{\lambda_e} \tag{4.44}$$

where $\Delta\lambda$ is the change in λ due to the magnetic field. A plot of $-B^2/\Delta\lambda$ against B^2 is a straight line with a slope of $1/\lambda_e$.

A similar relationship can be written for the Righi-Leduc coefficient S_{RL}:

$$-\frac{1}{S_{RL}} = \frac{1}{(S_{RL})_0} + \frac{R_H \lambda_L}{L_0 T} B^2 \tag{4.45}$$

R_H is the Hall coefficient and $(S_{RL})_0$ is the zero-field value of S_{RL}. A plot of $-1/S_{RL}$ versus B^2 is linear with a slope $R_H \lambda_L / L_0 T$.

Equations 4.44 and 4.45 are plotted for a number of samples in Figs. 4.7 and 4.8. The consistency of the results obtained is apparent and the electronic and lattice contributions to the thermal conductivity can evidently be obtained satisfactorily by these methods.

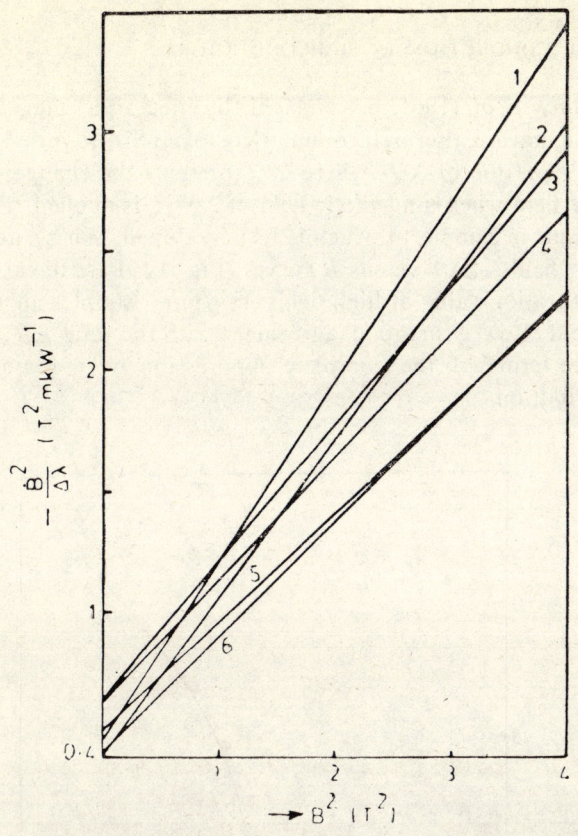

Fig. 4.7 $-B^2/\Delta\lambda$ plotted against B^2 for different samples of Cd_3As_2 at 300 K (after Armitage and Goldsmid, 1969). Curves refer to samples with carrier concentrations (10^{24} m^{-3}); Curve 1, 1.01; 2, 1.20; 3, 1.76; 4, 2.55; 5, 2.38; 6, 2.05.

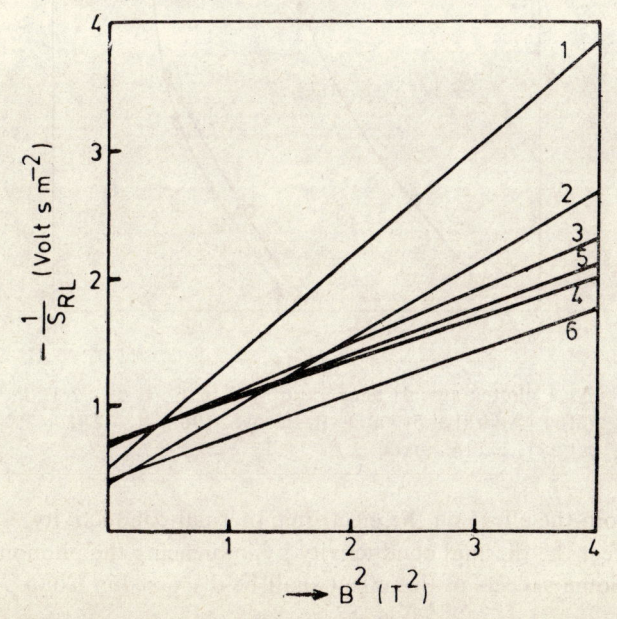

Fig. 4.8 Righi-Leduc coefficient $1/S_{RL}$ versus B^2 in Cd_3As_2 (after Armitage and Goldsmid, 1969). Curves refer to carrier concentrations as in Fig. 4.7.

Amirkhanov *et al.* (1961) investigated the effect of a magnetic field on the high-temperature thermal conductivity of InSb. Figure 4.9 shows the temperature variation of $\Delta\lambda/\lambda$ where $\Delta\lambda$ represents the change in λ due to the magnetic field when a magnetic field of 2.37 T is applied. The change is more significant in sample 14 which is heavily doped. These authors show that, at low fields, $\Delta\lambda/\lambda$ versus B curves (Fig. 4.10) are linear and tend to acquire a saturation value at high fields. For pure samples in a saturating magnetic field $\Delta\lambda/\lambda$ is in good agreement with the ratio λ_e/λ. This can be interpreted in terms of the complete suppression of λ_e leaving only the lattice contribution.

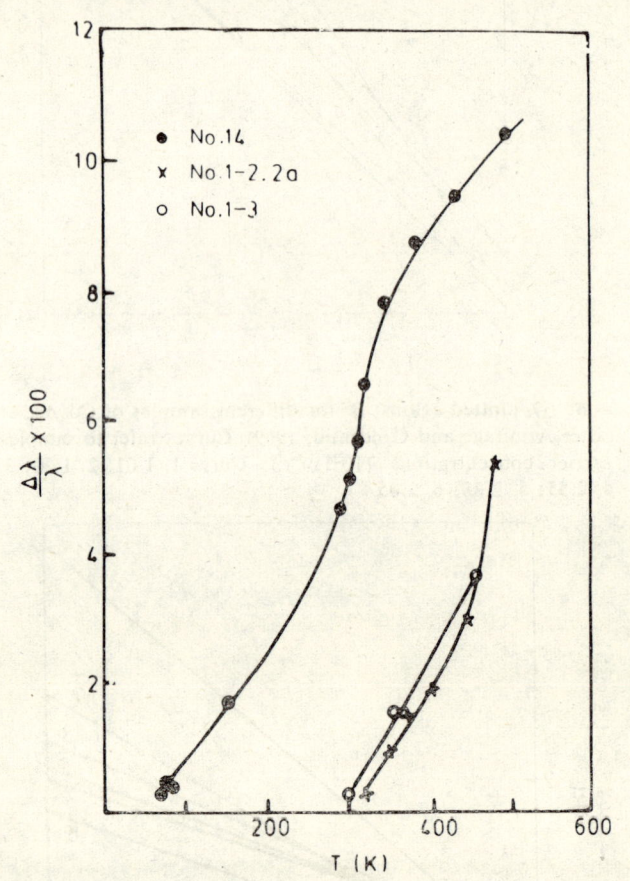

Fig. 4.9 $\Delta\lambda/\lambda$ plotted against temperature for InSb at $B = 2.37$ T (after Amirkhanov and Bashirov, 1961) $1-2$, $1-3$, p-type; $2a$, 14, n-type.

Apart from the effect on the electronic thermal conductivity, a magnetic field can affect the thermal conductivity by influencing the phonon-electron scattering. Some aspects of this effect shall be discussed in Chap. VIII.

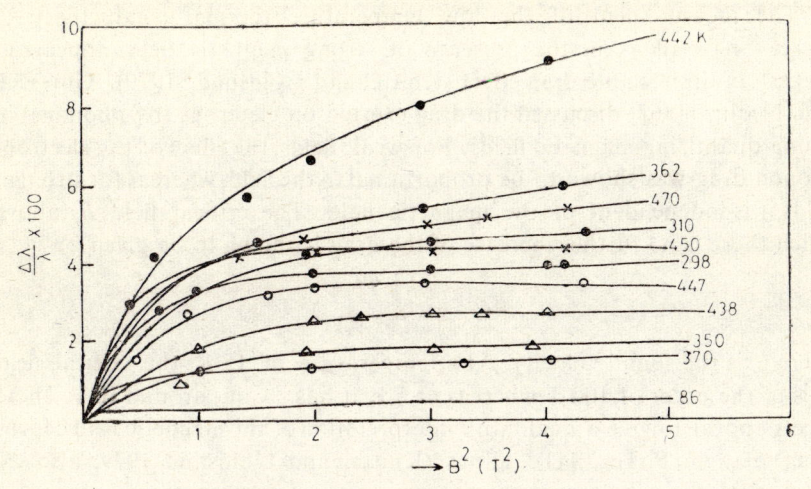

Fig. 4.10 $\Delta\lambda/\lambda$ versus B^2 at various temperatures in InSb (after Amirkhanov and Bashirov, 1961): ●, No. 14; x, No. 2a; o, No. 1 − 3; △, No. 1 − 2.

Appendix A

Electron-phonon drag

In the theory of heat transport by phonons electrons play the role of the scattering centres whereas they themselves are assumed to be in equilibrium. Similarly, the phonon system is assumed to be in equilibrium while describing electronic transport although they act as scattering centres for electrons. These assumptions are usually justified but there are situations when they are no more valid and a proper account has to be taken of the resulting 'phonon-drag' on electrons and 'electron-drag' on phonons. Gurevich (1945) considered the effect of phonon-drag on the thermoelectric effects in metals.

Parrott (1957) discussed the contribution of phonon-drag to thermal conductivity. The thermal conductivity of semiconductors which is made up of contributions from phonons and electrons is likely to be affected by the electron-phonon drag. For low carrier concentrations, the phonon contribution dominates and the change in the electron distribution produced by the phonon flow is expected to be very small. In heavily doped semiconductors, these effects are more likely to be observed. The electronic thermal conductivity can be written as $\lambda_e + \lambda_p$, where λ_e is the usual electronic contribution and λ_p the electron-drag term. The Lorenz number defined by $(\lambda_e + \lambda_p)/\sigma T$ may have quite large values at low temperatures. In n-type Ge at 80 K the calculated value of the Lorenz number is about 16 times the normal free electron value (Parrott, 1957). However, the electronic thermal conductivity is still small compared to the lattice contribution λ_L and any experimental verification is therefore difficult. No definite confirmation of a phonon-drag effect on the thermal conductivity has been reported.

The measurement of the low temperature thermal conductivity of PbTe—SnTe alloys in the presence of strong magnetic fields appears to suggest an appreciable drag effect (Knittel and Goldsmid, 1979). Gurevich and Nedlin (1962) discussed the drag exerted on electrons (by phonons) in strong quantizing magnetic fields. For weak fields, the effect of the electron-phonon drag was shown to be proportional to the field, whereas for stronger fields it is independent of the magnetic field. The critical field B_s above which there is no further increase of the drag is shown to be given by

$$B_s = \frac{k_B^2 \, T^2}{(2e \, \hbar^2 \, v_s^2)}$$

where v_s is the sound velocity. At a temperature of 15 K the critical field B_s is of the order of 100 T whereas at 5 K it has a value around 10 T. These effects appear to give a qualitative interpretation of the magnetic field dependence of λ in PbTe—SnTe alloys (Knittel and Goldsmid 1979, also see Chap. VIII).

References

Amirkhanov, D.Kh and Bashirov, R.I. (1961), *Sov. Phys.—Solid State 2*, 1447.

Armitage, D. and Goldsmid, H J. (1969), *J. Phys. C. 2*, 2138.

Bhandari, C.M. and Rowe, D.M. (1984), *9th European Thermophysical Properties Conference*, Manchester, September 1984.

Bhandari, C.M. and Rowe, D.M. (1985), *J. Phys. D : Appl. Phys. 18*, 873.

Drabble, J.R. and Goldsmid, H.J. (1961), *Thermal Conduction in Semiconductors*, Pergamon Press, London, Ch. 5.

Fistul, V.I. (1969), *Heavily Doped Semiconductors*, Plenum, New York.

Fulkerson, W., Moore, J.P., Williams, R.K., Graves, R.S. and McElroy, D.L. (1968), *Phys. Rev. 167*, 765.

Goldsmid, H.J. (1958), *Proc. Phys. Soc. 72*, 17.

Goldsmid, H J. (1965), *Materials used in Semiconductor Devices*, (ed. C.A. Hogarth), John Wiley and Sons Ltd, Interscience Publishers, London, p. 165.

Gurevich, L.E (1945), *J. Phys.*, Moscow 9, 477.

Gurevich, L.E. and Nedlin, G.M. (1962), *Sov. Phys.—Sol. State 3*, 2029.

Herrings, C. (1955), *Bell Syst. Tech. J. 34*, 237.

Kane, E.O. (1957), *J. Phys Chem. Solids 1*, 249.

Korenblit, L.L. and Sherstobitov, V.E. (1968), *Sov. Phys.—Semiconductors 2*, 573.

Knittel, T. and Goldsmid, H.J. (1979), *J. Phys. C: Solid St. Phys. 12*, 1891.

Parrott, J.E. (1957), *Proc. Phys. Soc., Lond. 70*, 590.

Parrott, J.E. (1979), *Rev. Int. Hautes Temper. Refract. Fr. 16*, 393.

Price, P.J. (1955), *Phil. Mag. 46*, 1252.

Ravich, Yu. I., Efimova, B.A. and Smirnov, I.A. (1970), *Semiconducting Lead Chalcogenides* (ed. L.S. Stil, bans), Plenum Press, New York.

Ravich, Yu. I., Efimova, B.A. and Tamarchenko, V.I. (1971), *Physica Stat. Solidi (b) 43*, 11.

Rowe, D.M. and Bhandari, C.M. (1983), *Modern, Thermoelectrics*, Holt Saunders, London.

Seeger, K. (1982), "*Semiconductor Physics—an Introduction*", Springer Verlag, Berlin.
Smirnov, I.A. and Ravich, Yu. I. (1967), *Sov. Phys.—Semiconductors 1*, 739.
White, G.K. and Woods, S.B. (1958), *Phil. Mag. 3*, 342.
Wilson, A.H. (1953), *The Theory of Metals*, Cambridge University Press.
Wright, D.A. (1970), *Metallurgical Reviews 15*, 147.
Zawadzki, W. (1974), *Adv. in Phys. 23*, 437.
Ziman, J.M. (1960), *Electrons and Phonons*, Clarendon Press, Oxford.

Chapter V

Phonons and Phonon Scattering Processe

5.1 Introduction

A crystalline solid is characterized by a periodic arrangement of atoms in a regular fashion. Interatomic forces are responsible for binding the atoms together and a crystal potential energy, which acquires a minimum value at the equilibrium separation of the atoms, can be defined. The atoms in a solid cannot, in general, leave their positions. However, they can perform vibrational motions about their mean positions. For a system consisting of a large number of atoms, the vibrational motion is complicated but can be analysed in terms of other less complex modes of vibration, referred to as normal modes. Each normal mode has a characteristic frequency, and in a crystalline solid these modes can be described either as standing or travelling waves.

5.2 Vibrations in a one-dimensional crystal

5.2.1 *One atom per unit cell*

Consider a linear chain of N identical atoms each of mass M and separated by a distance a. The force between two adjacent atoms is assumed to be proportional to the displacement of the atom from the mean position (so-called harmonic approximation, β is taken to be the force constant). The possible modes in which atoms can vibrate with angular frequency ω are described by

$$\omega(q) = \sqrt{(\beta/M)}\, 2\,|\sin(qa/2)| \qquad (5.1)$$

where q describes the wave number. For a three-dimensional solid, the analysis is much more complicated but, in general, a relationship of the type

$$\omega = \omega_s(\mathbf{q}) \qquad (5.2)$$

exists between the frequency of the mode and its wave vector $\mathbf{q}$. The subscript s refers to the polarisation of the wave.

The wave vector $\mathbf{q}$ has a magnitude equal to $2\pi/\lambda$, where λ is the wavelength associated with a particular mode. The direction of $\mathbf{q}$ is the direction in which the wave propagates. The displacement of the atom can be described in terms of its two components: one more or less parallel to the direction of propagation (longitudinal mode) and the other more or less perpendicular to this direction (transverse mode).

The case with only one type of atom in a linear chain is a simple one. There is one atom per unit cell and the ω-q relationship is as described in Fig. 5.1 (a). At low frequencies there is a linear variation of ω with q and one can describe a sound wave with velocity v_s defined by

$$\omega = v_s q \tag{5.3}$$

At higher frequencies the linear ralationship does not hold and this gives rise to dispersion. The resulting dependence of ω on q is referred to as a dispersion relation. It must be noted that q is restricted to the region $-\pi/a < q < \pi/a$, and the slope of the ω-q curve becomes zero at $q = \pm \pi/a$. In three dimensions **q** is confined to a certain volume in **q**-space which is referred to as the Brillouin zone.

5.2.2 *Two atoms per unit cell – n atoms per unit cell*

Let us now consider the linear chain with two atoms per unit cell and with two types of atoms having masses M_1 and M_2. There will be two nearest-neighbour force constants involved, β_1 between atoms in the same cell and β_2 between the nearest atoms in adjacent cells. Of the two solutions obtained for the equations of motion one is the usual acoustic mode as ω approaches zero with q approaching zero. The second solution refers to the optic mode and here ω remains finite as q goes to zero. The name optic mode is given due to the frequency being in the infra-red region. The dispersion curves for a diatomic linear chain are displayed in Figs. 5.1 (b) and 5.1 (c).

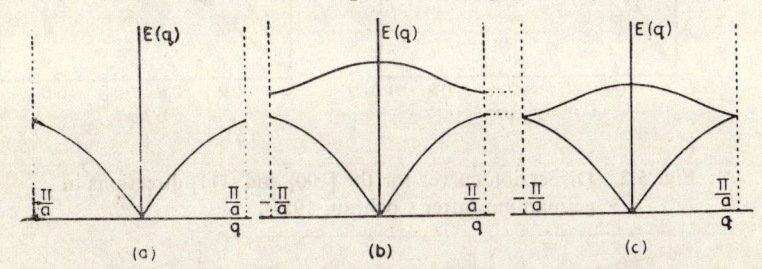

Fig. 5.1 $E(\mathbf{q})$ versus **q** curves for; (a) monatomic linear chain, (b) diatomic linear chain with $M_1 \neq M_2$ and (c) diatomic linear chain with $M_1 = M_2$.

5.3 Real tcrysals

5.3.1 *Theoretical considerations*

For a real crystal in three dimensions, a similar approach can be followed provided that we have some knowledge of the interatomic forces. A three-dimensional cubic crystal with two atoms per unit cell has three acoustic (2 transverse and 1 longitudinal) branches and three optic branches. For certain directions of propagation, the two transverse branches may coincide. If there are n atoms per unit cell, there will be three acoustic

80 THERMAL CONDUCTION IN SEMICONDUCTORS

branches and $(3n - 3)$ optic branches. A detailed review of the theoretical aspects of the subject has been given by Maradudin *et al* (1963).

A calculation of the $E(q)$ versus q curves requires assumptions about the interatomic forces. Different approximations are required for different types of crystals which lead to different forms of lattice dynamics, such as in the rigid-ion model, the point-polarizable model, the shell model and related approximations.

In Figs. 5.2 and 5.3 are displayed the dispersion curves for germanium (Cochran, 1959) and silicon (Dolling, 1963) respectively.

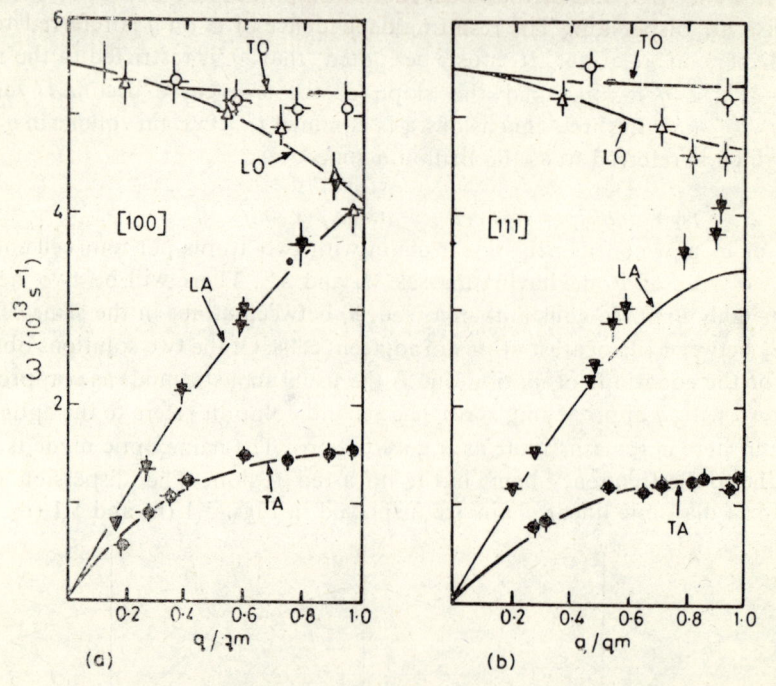

Fig. 5.2 Dispersion curves for the [100] and [111] directions in germanium (after Cochran, 1959).

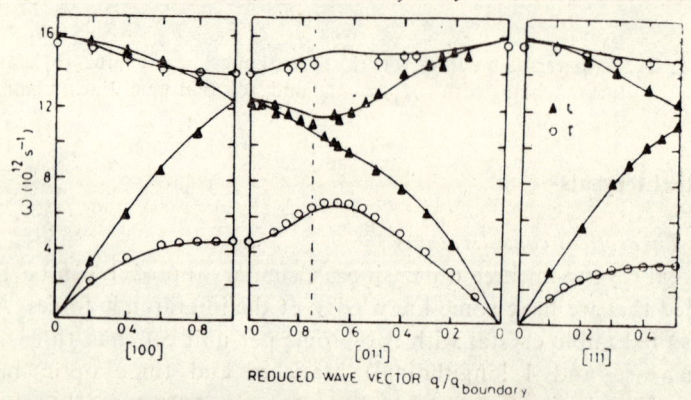

Fig. 5.3. Phonon dispersion curves in silicon at 296 K in three principal directions (after Dolling, 1963).

5.3.2 *Phonons*

The analysis of the vibrational modes of a large number of atoms in a three-dimensional solid is complicated. The complex vibrational modes can be simplified by analysis in terms of normal modes. A normal mode has a characteristic frequency associated with it and the corresponding atomic movements can occur completely independently of any other normal mode. The energy in a mode of frequency ω can have values $(n + \frac{1}{2})\hbar\omega$, where n is any positive integer. The zero-point energy term given by $\frac{1}{2}\hbar\omega$ is not important in the present context and the remaining energy is referred to as the energy of n phonons, each having energy $\hbar\omega$. The phonons are the quanta of excitations of the normal modes of lattice vibration; the analogy has been derived from the photon as a quantum of electromagnetic wave.

The concept of phonon has been extremely useful in analysing the vibrational motion of a very large number of atoms.

A phonon has associated with it a momentum $\hbar q$ and energy $\hbar\omega$. Phonons have no mass and their number is not conserved, unlike in the case of atoms or electrons.

Density-of-states

A phonon density-of-states function can be defined such that $g(\nu)\,d\nu$ is the fraction of the phonon modes in the frequency range ν and $\nu + d\nu$. Computation of $g(\nu)$ requires a complete solution of the dynamical equation for the lattice modes. Very often the simple Debye model is used where the density-of-states $g(\nu)$ varies as ν^2 up to a cut-off, and is zero for higher frequencies. In the Einstein model $g(\nu) \sim \delta(\nu - \nu_E)$. More realistic calculations of $g(\nu)$ show that a $g(\nu)$ versus ν plot exhibits a wide spread with several peaks which correspond to the modes of different polarizations having very different velocities. The low-frequency part of the spectrum has a close resemblance to the Debye spectrum.

5.3.3 *Experimental methods*

A number of experimental methods are employed in obtaining information about phonon dispersion relations. X-rays and neutrons can be scattered inelastically by crystals and these scattering experiments can be used to obtain phonon dispersion relations. These experiments also provide evidence for the existence of phonons, as in the scattering process the changes in energy and momentum correspond to the creation or absorption of one phonon.

In non-metallic crystals, optical absorption by impurities sometimes shows a structure of multiple spikes along with the main absorption line. These discrete spikes may be interpreted as arising from the emission of a number of phonons. The interactions of phonons with x-rays, neutrons and photons have been reviewed by Bak (1964) and Stevenson (1966).

Inelastic scattering of neutrons by phonons

A neutron sees the crystal lattice chiefly by nuclear interactions that take place between the incident neutron and atomic nuclei. If the wavevectors of incident and scattered neutrons are represented by $\mathbf{K}$ and $\mathbf{K}'$ and if $\mathbf{q}$ is the wavevector of the phonon created (or absorbed) the general wavevector conservation relation is given by

$$\mathbf{K} = \mathbf{K}' + \mathbf{q} + \mathbf{G} \tag{5.4}$$

$\mathbf{G}$ is a reciprocal lattice vector. The incident neutron of mass M_n has a kinetic energy $p^2/2M_n$ and a momentum $\mathbf{p} = \hbar \mathbf{K}$. The energy conservation requirement gives

$$\hbar^2 K^2 / 2M_n = \hbar^2 K'^2 / 2M_n \pm \hbar \omega_q \tag{5.5}$$

ω_q is the frequency of the phonon. The scattering angle is given by $(\mathbf{K} - \mathbf{K}')$. In an experiment it is necessary to determine the energy gain or loss of the scattered neutrons as a function of $(\mathbf{K} - \mathbf{K}')$ (Brockhouse, 1964 and 1966). Under favourable conditions, and with large good single crystals, this is the ideal method for the determination of phonon dispersion relations. It is also possible to obtain information on phonon life times from the width of the scattered beam.

5.4 Effect of defects—lattice imperfections

The introduction of an impurity atom into a crystal lattice brings about changes in the vibrational spectrum. The normal modes of the crystal may change depending upon the nature of the impurity. In addition, new modes may appear in the frequency ranges forbidden to the normal modes of the host crystal. A light impurity atom gives rise to a mode of vibration whose frequency is greater than the maximum frequency for the host lattice. Such modes of vibration are referred to as localized modes. Sometimes, new modes appear in the gap between the acoustic and optic phonon branches; these modes are called gap modes. In KCl, the replacement of Cl$^-$ by the I$^-$ impurity gives rise to such modes. New modes may also appear in regions allowed to the normal modes of the host crystal. This happens for certain kinds of impurities, particularly heavy impurities. Replacement of K$^+$ in KI by Ag$^+$ gives rise to this type of modes and these are referred to as resonant modes.

In all these cases, the introduction of impurities can have a profound effect on a number of observable properties. The direct observation of localized modes in the infrared absorption spectra of alkali halide crystals containing hydride ion impurities substituting the halogen ions (Schaefer, 1960), incoherent neutron scattering in Ni-Pd alloys (Mozer et al, 1962) and the thermal conductivity of alkali halide crystals containing impurities (Pohl, 1962) have been instrunental in bringing about a renewed interest in the study of the vibrational properties of imperfect lattices. Maradudin

(1966) has given a detailed account of the subject. In the present context we shall be interested mainly in the aspect of defect behaviour which influences the phonon mean-free-path by various phonon scattering processes. The presence of new modes, such as resonant modes, also has its effect on thermal conductivity. This will be considered in Sec. 5.9.

5.5 Vibrational properties of non-crystalline solids

A long-range order exists in the vibrational properties of solids with or without defects discussed so far. Also, whatever be the degree of disorder introduced by impurities, the basic features of vibrational spectrum of a completely ordered solid are retained. In non-crystalline materials, such a long-range order does not exist and various experiments suggest that the vibrational spectrum is primarily governed by a short-range order. The effect of the absence of a long-range order is to smear out the details of the vibrational density-of-states curves and to give rise to an enhancement of the number of low-frequency modes. Inelastic coherent neutron-scattering experiments suggest that the low-frequency excitations can be interpreted in terms of highly damped phonon-like modes giving rise to one or two acoustic branches.

Theoretical methods used in the investigation of the vibrational properties of non-crystalline solids have been reviewed by Hori (1968), Bell (1972) and Dean (1972). Bottger (1974) has reviewed various aspects of the experimental and theoretical work on disordered solids.

5.6 Phonon Boltzmann equation—relaxation time

The problem of obtaining an expression for lattice thermal conductivity is essentially the problem of obtaining the phonon distribution function N_q. In thermal equilibrium the average number of phonons of energy $\hbar\omega$ is

$$N^0 = \{\exp(\hbar\omega/k_B T) - 1\}^{-1} \qquad (5.6)$$

For $\hbar\omega/k_B T \ll 1$, $N^0 = k_B T/\hbar\omega$.

Since $\hbar\omega$ is the energy of the phonon, the average energy for the normal mode is $k_B T$.

A phonon of energy $\hbar\omega$ and a velocity $\mathbf{v}_s(\mathbf{q})$ in the direction of $\mathbf{q}$ contributes $\hbar\,\omega_s(q)\,\mathbf{v}_s(\mathbf{q})$ to the heat current. The net heat current is then given by

$$\mathbf{h} = \sum_{\mathbf{q},s} N_{q,s}\,\hbar\,\omega_s(\mathbf{q})\,\mathbf{v}_s(\mathbf{q}) \qquad (5.7)$$

The index s refers to the phonon polarization. In equilibrium

$$N_{\mathbf{q},s} = N^0_{\mathbf{q},s}$$

$$\omega_s(\mathbf{q}) = \omega_s(-\mathbf{q})$$

Further, the group velocity $\mathbf{v}_s(\mathbf{q})$ is equal and opposite to $\mathbf{v}_s(-\mathbf{q})$. As the index s does not directly influence the development of the theoretical formulation it can be omitted at this stage and will be reintroduced later in Chap. VI.

Obviously, $\mathbf{h} = 0$ in thermal equilibrium. There would be a net heat current only if $N(\mathbf{q})$ departs from its equilibrium value $N^0(\mathbf{q})$. $N(\mathbf{q})$ may change because of an external field or a temperature gradient. Assuming a temperature gradient along the x-direction we can write

$$(\partial N_\mathbf{q}/\partial t)_{\text{drift}} = -V_x(\partial N_\mathbf{q}/\partial x) = -V_x(\partial N_\mathbf{q}/\partial T)(\partial T/\partial x) \qquad (5.8)$$

The corresponding three-dimensional equation can be written as

$$(\partial N_\mathbf{q}/\partial t)_{\text{drift}} = -(\mathbf{V} \cdot \nabla T)\frac{\partial N_\mathbf{q}}{\partial T} \qquad (5.9)$$

However, in the steady-state the phonon density at any point must become independent of time.

There must, therefore, exist processes which tend to oppose the density change due to the drift of phonons and help bring the distribution to a steady state. Phonons may encounter resistance due to various processes, such as scattering by other phonons, impurities, charge-carriers, grain-boundaries, etc. The change in the phonon distribution due to all such processes can be written as $(\partial N_q/\partial t)_{\text{scatt}}$ and in the steady state

$$(\partial N_\mathbf{q}/\partial t)_{\text{drift}} + (\partial N_\mathbf{q}/\partial t)_{\text{scatt}} = 0 \qquad (5.10)$$

This is the Boltzmann equation for phonons similar to the one described (in Chap. III) for electrons.

There are two different approaches to solve the Boltzmann equation (See Chaps. III and VI). The essential steps in obtaining a solution of the phonon Boltzmann equation are similar to those used in the case of electrons. In the relaxation-time approximation, the rate of change of the phonon distribution due to scattering processes is written as

$$(\partial N_\mathbf{q}/\partial t)_{\text{scatt}} = (N_\mathbf{q}^0 - N_\mathbf{q})/\tau(\mathbf{q}) \qquad (5.11)$$

One then has a well-defined relaxation time $\tau(\mathbf{q})$. In the case of phonons, it is not always possible to derive an expression for a well-defined relaxation time. In this situation, a second approach—the variational approach—is more useful. In this method, the different scattering mechanisms are incorporated in the form of transition probabilities which can then be calculated by using the perturbation theory. Both the methods have their own merits and demerits. However, inspite of an initial interest in the variational approach, most of the effort towards the understanding of the thermal transport in solids has been based on the relaxation-time method.

5.7 Scattering of phonons

Under the influence of a temperature gradient, the phonon distribution deviates from its equilibrium value and a heat current is produced. A steady flow of heat requires the presence of resistive processes as described in the previous section. This is the most important mechanism of heat transport in insulators and semiconductors. In heavily doped semiconductors, the electronic contribution to heat transport may become significant although most of the heat is still carried by phonons. Amongst the resistive processes which limit the phonon mean-free-path, the scattering of a phonon by other phonons is important. In a large perfect crystal, this is the sole mechanism which restricts the phonon mean-free-path and leads to a finite thermal conductivity. Other important mechanisms are the scattering of phonons by lattice disorder and by crystal boundaries. In doped semiconductors phonon-scattering by electrons (or holes) may constitute a significant part of the total scattering cross-section.

5.7.1 *Phonon-phonon interaction*

In a harmonic crystal, terms only up to the second order in the expansion of the crystal potential are retained. In such a system phonons would have an infinite lifetime. The introduction of third and higher order terms causes an interaction between phonons which now acquire a finite lifetime. The physical basis of phonon-phonon interaction lies in the anharmonic terms.

The most significant of these phonon interactions are the so-called three-phonon processes. In one such process, two phonons represented by wave vectors q_1 and q_2 combine to form a single phonon of wave vector q_3. In another situation one phonon breaks up and two phonons are created. It is convenient to represent these processes in a diagram where solid lines show the propagation of phonon and meeting of these lines corresponds to phonon-phonon interaction (Fig. 5.4).

5.7.2 *Normal and Umklapp processes*

It is expected that these processes will conserve energy and momentum. Energy conservation requires

$$\omega_1 + \omega_2 = \omega_3 \quad \text{for process shown in Fig. 5.4 (a)}$$
$$\omega_1 = \omega_2 + \omega_3 \quad \text{for process (b)} \tag{5.12}$$

and momentum conservation requires

$$q_1 + q_2 = q_3 \quad \text{for (a)}$$
$$q_1 = q_2 + q_3 \quad \text{for (b)} \tag{5.13}$$

However, not all processes conserve phonon momentum. In general a three-phonon interaction may be described by

$$q_1 + q_2 = q_3 + G \tag{5.14}$$

where G is a reciprocal lattice vector. For $G = 0$, the total crystal (or phonon) momentum is conserved and the process is referred to as a normal or N-process. Other values of G ($\neq 0$) correspond to umklapp or U-processes in which the momentum is not conserved. These processes are also referred to

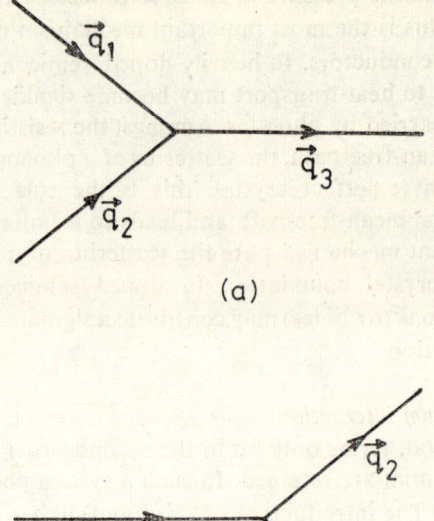

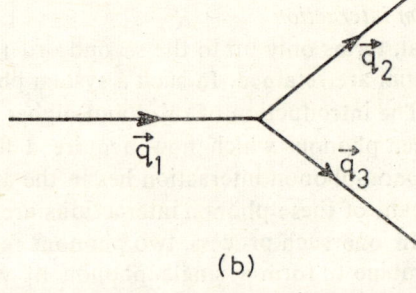

Fig. 5.4 Three-phonon processes: (a) two phonons of wavevectors q_1 and q_2 combine to create a phonon of wavevector q_3, and (b) a phonon of wavevector q_1 breaks up and two phonons are created.

as momentum destroying processes. Although G can take an infinitely large number of values, it is only the smallest reciprocal lattice vectors which appear in Eq. 5.14.

Normal or momentum conserving processes conserve the total phonon momentum. A phonon flow once started would continue even in the absence of a temperature gradient. By themselves normal processes cannot limit the phonon mean-free-path and, therefore, cannot produce any thermal resistance. Umklapp processes have to be taken into account to explain a finite mean-free-path for the phonons. The two types of processes are important in the establishment of thermal equilibrium in different ways.

For the N-processes described above, the three wavevectors must form a triangle (Fig. 5.5). Energy conservation requires, $\omega_1 + \omega_2 = \omega_3$ or (for small q) $v_1 q_1 + v_2 q_2 = v_3 q_3$; v_1, v_2 and v_3 are the phase velocities of the phonons participating in the process. For an isotropic solid with no dispersion, $v_1 = v_2 = v_3$ and therefore $|q_1| + |q_2| = |q_3|$. It is impossible to construct such a closed triangle. On this basis Peierls pointed out that the three interacting modes could not belong to the same polarization branch. The triangle can be closed only if v_3 is greater than either v_1 or v_2 or both. Therefore, processes allowed by these restrictions should be of the following type (Herpin, 1952).

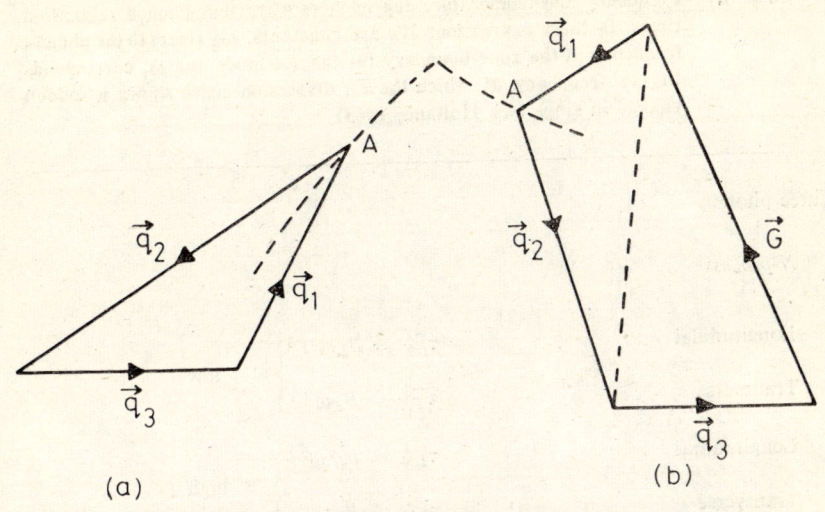

Fig. 5.5 The phonon modes participating in a three-phonon process: (a) N-process, (b) U-process. For a fixed q_3 each set is specified by a vertex A; the locus of all possible vertices is a surface of revolution about the axis q_3 or $q_3 + G$ in (a) and (b) (after Klemens, 1958).

$$\text{Transverse} + \text{Transverse} = \text{Longitudinal}$$
$$\text{Transverse} + \text{Longitudinal} = \text{Longitudinal}$$

For an Umklapp process $q_1 + q_2 = q_3 + G$, a quadrilateral diagram is constructed (Fig. 5.5 (b)) and it appears that there are more degrees of freedom in constructing this diagram than for the Normal processes (see Klemens, 1958).

5.8 Relaxation times for three-phonon processes

The three-phonon processes arise due to the anharmonic nature of the crystal potential energy and, therefore, a knowledge of the phonon spectrum is essential in the derivation of three-phonon relaxation times. In obtaining these relaxation times, Herring (1954) has explicitly taken the crystal class into account. Such expressions are therefore valid for this particular class of materials although these have often been used to analyse the experimental thermal conductivity data in other classes of materials.

Various three-phonon relaxation times have been obtained (Klemens 1951, 1958) without taking dispersion into account. For materials such as germanium and silicon the transverse acoustic (TA) branch shows a significant change in slope at around halfway towards the zone boundary. Holland (1963) has obtained phonon relaxation times for such cases. The dependences of the various three-phonon relaxation times on ω and T are shown in Table 5.1.

Table 5.1 Frequency and temperature dependences of various phonon relaxation times. In these expressions B's are constants. ω_2 refers to the phonon frequency at the zone boundary for the TA mode and ω_1 corresponds to the frequency at which the TA dispersion curve shows a sudden change in slope (see Holland, 1963)

Three phonon

  N-process[1]

Longitudinal	$\tau_{LN}^{-1} = B_L \omega^2 T^3$	low T
Transverse	$\tau_{TN}^{-1} = B_T \omega T^4$	
Longitudinal	$\tau_{LN}^{-1} = B_L' \omega^2 T$	high T
Transverse	$\tau_{TN}^{-1} = B_T' \omega T$	

  U-process

  Klemens[2] $\qquad \tau_U^{-1} = B_U \omega^2 T^3 \exp(-\theta/\alpha T)$

  Klemens[3] $\qquad \tau_U^{-1} = B_U \omega T^3 \exp(-\theta/\alpha T)$

  Dispersive transverse[4] $\quad \tau_{TU}^{-1} = B_{TU} \omega^2 / \sinh z, \quad \omega_1 < \omega < \omega_2$
  $\qquad\qquad\qquad\qquad\qquad\quad\; = 0, \qquad\qquad\qquad\; \omega < \omega_1$
  $\qquad\qquad\qquad\qquad\qquad\; z = \hbar\omega/k_B T$

  Callaway[5] $\qquad \tau_U^{-1} = B_0 \omega^2 T^3$

  Klemens[6] $\qquad \tau_U^{-1} = B' \omega^2 T \quad$ (High T)

Four phonon[7] $\qquad \tau_F^{-1} \propto \omega^2 T^2 / M \theta^2 v_s^3$

---

1. Herring (1954)
2. Klemens (1951)
3. Klemens (1958)
4. Holland (1963)
5. Callaway (1959)
6. Klemens (1958)
7. Pomeranchuk (1941)

Anisotropy may play an important role in producing phonon collision processes and thus influence the frequency dependence of the relaxation times (Herring, 1954). However, due to difficulties involved in working out suitable expressions for anisotropic materials, the relaxation times given in the Table 5.1 are frequently used in the data analysis.

The three-phonon processes considered usually involve only acoustic phonons but the processes where one of the phonons is optical are also possible. The type of interaction $A + A \rightleftharpoons 0$ (where A and 0 refer to acoustic and optic modes), can occur if the frequency of the optic phonon is not greater than twice the highest acoustic mode frequency. Such processes could be significant if the mass-ratio of atoms in the material does not deviate significantly from unity (Steigmeier and Kudman, 1966).

Collision processes involving four or more phonons arise from higher order anharmonicity of the interatomic forces. These processes are expected to be important at high temperatures (Joshi et al., 1970). However, there have been suggestions (Ecsedy and Klemens, 1975, and Klemens and Ecsedy, 1976) that the strength of four-phonon processes is very weak in comparison with that of the three-phonon processes. This subject is further discussed in Chap. VI, Appendix C.

5.9 Other phonon scattering mechanisms—scattering by defects in the lattice

Apart from phonon-phonon scattering, other important phonon-scattering mechanisms arise from the finite size of the specimen and various types of crystal imperfections. The effect of crystal boundaries on the phonon mean-free-path will be discussed in Chap. VI and the rest of the present section will be devoted to the scattering by various types of crystal imperfections.

A perfect large crystal would only have phonon-phonon interactions to limit the phonon mean-free-path. However, in all real crystals a number of irregularities of the lattice may exist, such as substitutional impurities, interstitials and dislocations. A phonon may be scattered by one or more of these imperfections causing a reduction in lattice thermal conductivity.

In a perfect large crystal (in the harmonic approximation), a phonon can move unhindered. The functional form of the crystal Hamiltonian is invariant under a transformation $\mathbf{X} \to \mathbf{X} + \mathbf{R}_n$, where $\mathbf{R}_n$ is a lattice translation vector. This invariance is destroyed by the presence of lattice imperfections and the resulting Hamiltonian can be expressed as $H = H_0 + H'$, where H_0 is still invariant with respect to a lattice translation but not H', which now acts as a perturbation Hamiltonian.

A method of analysing various imperfections is to divide them into different groups by their dimensionality, such as point imperfections, line imperfections, surfaces of imperfections and volume disorder. If the region over which an imperfection extends has linear dimensions much smaller than a phonon wavelength, the imperfection is said to be a point imperfection or a point defect. An isolated substitutional impurity, an interstitial atom, or a vacancy are common point imperfections. Dislocations are typical examples of line imperfections, whereas grain boundaries, twin boundaries and stacking faults are classified as surfaces of imperfections. The nature of imperfection in substitutional alloys and glasses can be referred to as volume disorder (for details, see Ziman, 1960).

5.9.1 Mass-difference scattering

For point imperfections the scattering of phonons may be caused by the differences in mass, or the force constant. Pomeranchuk (1942) pointed out that fluctuations in mass-difference throughout the crystal cause thermal resistance. Klemens (1955) obtained an expression for thermal resistance due to mass-difference scattering. The phonon relaxation time for this case can be obtained in a simple form

$$\tau_d^{-1} = \frac{\Gamma \omega^2 g(\omega)}{6N} \quad (5.15)$$

Γ, which measures the strength of scattering, is given by

$$\Gamma = \sum_i f_i \left(1 - \frac{M_i}{\overline{M}}\right)^2 \quad (5.16)$$

f_i is the fractional concentration of the impurity atoms of mass M_i and $\overline{M} = \sum_i f_i M_i$, is the average atomic mass, For the Debye model with $g(\omega) = 3V\omega^2/2\pi^2 v_s^3$ (V is the crystal volume and v_s the average sound velocity), the familiar Rayleigh expression is obtained

$$\tau_d^{-1} = \frac{\Omega_0 \Gamma \omega^4}{4\pi v_s^3} \quad (5.17)$$

where $\Omega_0 = V/N$, is the average atomic volume.

5.9.2 More on disorder scattering

In general, an impurity atom differs from the host atoms in its mass, its size and in the force constants. Moreover, as a result of anharmonicity the force constants are modified by the 'misfit' strain-field in the neighbourhood of the impurity. Klemens (1955) analysed point defect scattering by an impurity in a simple cubic lattice. The relaxation time obtained by Klemens is similar to the one derived for the mass-difference scattering discussed earlier but with the parameter Γ now incorporating differences in volume and stiffness constants (see Carruthers, 1961, and Abeles, 1963).

$$\Gamma_i = f_i \left[(\Delta M_i/M)^2 + 2\{(\Delta G_i/G) - 6.4\gamma (\Delta \delta_i/\delta)\}^2\right] \quad (5.18)$$

G_i is the average stiffness constant of the nearest-neighbour bonds from impurity to host lattice and G is the corresponding quantity for the host atoms. δ_i' is the cube root of atomic volume of the i-th impurity in its own lattice and $\delta = \sum_i f_i \delta_i'$. γ represents an average anharmonicity of the bonds.

G can be identified with the bulk modulus and $\Delta G/G$ can be expressed in terms of $\Delta \delta/\delta$ with the help of the following relationship (Keyes, 1962)

$$C_{ik}\delta^4 = \text{constant} \quad (5.19)$$

where C_{ik} are the elastic constants. The constant in the above equation acquires a different value for different crystal groups. Using these relations, it follows that

$$\Gamma_i = f_i\,[(\Delta M_i/M)^2 + \epsilon\{(\delta - \bar{\delta}_i)/\delta\}^2] \qquad (5.20)$$

where ϵ has a value around 39; it is sometimes treated as an adjustable parameter.

However, it must be mentioned that it is not justified to add up the squares of the relative change in atomic mass and the relative change in the force constant to obtain the parameter Γ. The relative changes in atomic mass and force constant should first be added and then squared as the two scattering processes may tend to cancel each other in some situations.

However, of all the contributions to Γ only the one from mass-difference scattering is well understood. Other terms have been mostly used to obtain an estimate of the total scattering by the impurity, rather than to seek a quantitative agreement with the observed data. For isotopic defects, Berman and Brock (1965) reported a good experimental verification of this theory.

Several workers have described more refined theories of the defect scattering for arbitrary mass and force constant of the impurity atom (Callaway, 1963, Klein, 1963 and 1966, Krumhansl, 1963, Takeno, 1963, and Yussouf and Mahanty, 1965 and 1966).

It was observed that the introduction of KNO_2 and KI impurities in KCl gives rise to a dip on the low-temperature side of the maximum in the $\lambda-T$ curve, whereas for monatomic impurities the dip was found to occur on the high-temperature side of the conductivity maximum. Pohl (1962) pointed out that the $KCl - KNO_2$ dip could be reproduced by a resonance scattering of phonons which may arise due to their inelastic scattering at the impurity modes. Holland and Neuringer (1962) and Vook (1965) discussed this type of scattering in semiconductors. The dip observed in the $\lambda - T$ curve can be reproduced (Walker and Pohl, 1963) by using an additional relaxation time given by Wagner (1963)

$$\tau_R^{-1}(\omega) = \frac{c_1 T^n \omega^2}{(\omega_0^2 - \omega^2)^2 + c_2^2 \omega_0^2 \omega} \qquad (5.21)$$

c_1 and c_2 are constants and ω_0 is a frequency characteristic of the impurity mode. For monatomic disturbances {such as in KCl: $NaCl$}, $n = 2$ gives a good agreement, whereas for polyatomic impurities (as in KCl: KNO_2), $n=0$ is found to be more suitable. This type of resonance scattering may be due in many cases to the inelastic scattering of phonons interacting with the localized modes (Wagner, 1963 and 1964).

5.9.3 *Scattering by dislocations*

In some materials, scattering of phonons by dislocations may play an important role in limiting the phonon mean-free-path. The core of the dislocation constitutes a region of disorder and is expected to scatter phonons according to a relaxation time $\tau_R^{-1} \propto \omega^3$. However, the results of Sproull et al (1959) show that the thermal resistance can be larger than predicted by the initial calculations of the dislocation scattering by Klemens (1955) since the strain-field associated with the dislocation may cause a significant phonon scattering. Klemens included this effect in his subsequent calculations which were later refined by other workers. The anharmonic effect of the strain-field can be specified by the Gruneisen constant γ. For the case of randomly arranged screw dislocations, the phonon relaxation time is given by

$$\tau_{\text{str.}}^{-1} = \frac{2^{3/2} N_d b^2 \gamma^2 \omega}{27(3^{1/2} + 2^{1/2})} \qquad (5.22)$$

N_d is the number of dislocation lines through a unit area and b is the magnitude of the Burgers' vector. For edge dislocations a similar expression is obtained with an additional term

$$\left[1/2 + \frac{\beta^2}{24}[(1 + \sqrt{2}(v_L/v_T)^2)^2] \right]$$

where $\beta = (1 - 2\nu)/(1 - \nu)$. Here ν is Poisson's ratio, and v_L and v_T refer to the longitudinal and transverse phonon velocities.

Similar expressions were obtained by Nabarro (1951), Ziman (1960), Carruthers (1961), Bross et al (1963) and Ohashi (1968), and these differ mostly in the numerical constant. In several experimental studies of the dislocation scattering of phonons in non metallic crystals, the actual scattering was found to be considerably larger than that obtained by considering fixed dislocations. The effect of vibrating dislocations was considered by Ishioka and Suzuki (1963), and the experiments of Anderson and Malinowski (1972) and Suzuki and Suzuki (1972) provided evidence of the effect of this type of scattering on thermal conductivity.

5.10 Scattering of phonons by electrons (or holes)

5.10.1 *Introduction*

In metals, the lattice thermal conductivity is small compared to the electronic contribution due to a considerable scattering of phonons by conduction electrons. In semiconductors, the concentration of free charge carriers (electrons or holes) can be adjusted by suitable doping, and the ratio of electronic and lattice thermal conductivities varies with the level of doping. The lattice thermal conductivity in general decreases with increasing carrier concentration and it is reasonable to believe that this reduction in the lattice contribution may arise due to the scattering of phonons by electrons or holes. The same interaction between electrons and phonons also causes the scattering of electrons and gives rise to an electrical resistance (see Secs. 3.10 and 7.3). The net effect of an increase in carrier concentration may be to increase the total thermal conductivity as a result of the increase in the electronic component.

5.10.2 *Scattering by free carriers*

At high temperatures, donor atoms are ionized and phonons are scattered by electrons in the conduction band. However, at low temperatures a significant fraction of electrons may still be trapped by donors, and in that case scattering of phonons by electrons bound to donors has to be taken into account. Both types of cases have been discussed in literature and, although a qualitative understanding of the effect of phonon scattering by electrons can be reasonably made on these lines, a quantitative analysis of the observed thermal conductivity may be difficult when both types of processes are operative.

Ziman (1956) obtained an expression of the phonon relaxation time for scattering by free electrons in a parabolic band. The derivation based on the effective-mass-approximation gives

$$\tau_{pe}^{-1} = \frac{\epsilon_1^2 m^{*2} k_B T}{2\pi \hbar^4 \rho v_L} \left\{ z - \ln \left[\frac{1 + \exp\left(\frac{\hbar^2 \omega^2}{8m^* v_L^2 k_B T} + \frac{m^* v_L^2}{2k_B T} - \frac{E_F}{k_B T} + \frac{z}{2}\right)}{1 + \exp\left(\frac{\hbar^2 \omega^2}{8m^* v^L k_B T} + \frac{m^* v_L^2}{2k_B T} - \frac{E_F}{k_B T} - \frac{z}{2}\right)} \right] \right\}$$

(5.23)

where $z = \hbar\omega/k_B T$, ρ is the density and E_F the Fermi energy. This relaxation time was derived for longitudinal phonons although a similar relaxation time is also taken to be valid for transverse phonons. Here ϵ_1 specifies the strength of the electron-phonon interaction and is referred to as the deformation potential. v_L is the longitudinal phonon velocity and m^* the electron effective-mass.

For long-wavelength phonons, the expression can be simplified further. At low carrier concentration Eq. 5.23 can be written as (Parrott, 1979)

$$\tau_{pe}^{-1} = \frac{n \epsilon_1^2 \omega}{\rho v_L^2 k_B T} \sqrt{\frac{\pi m^* v_L^2}{2 k_B T}} \, \exp\left(-m^* v_L^2 / 2 k_B T\right) \quad (5.24)$$

n is the carrier concentration. For large carrier concentrations, one obtains

$$\tau_{pe}^{-1} = \frac{\epsilon_1^2 m^{*2} \omega}{2\pi \hbar^3 \rho v_L} \quad (5.25)$$

In either case τ_{pe}^{-1} is proportional to ω. In the second case the proportionality constant is independent of n. A plot of the function $f(x)$ which appears in τ_{pe} against $x(= \omega/\omega_D)$ is shown in Fig. 5.6 for $\xi(= E_F/k_B T) = 4$ (Parrott, 1979).

5.10.3 *Scattering by electrons in bound states*

The lattice thermal conductivity (λ_L) of doped semiconductors shows a number of interesting features. There is substantial reduction in λ_L below the conductivity maximum even for dilute concentrations. Moreover, λ_L versus temperature curve sometimes shows a variation faster then T^3. In some cases the curves may show dips.

Keyes (1961) proposed the following explanation for the scattering of phonons by bound electrons. He analysed the case of germanium. The ground state of a donor in germanium is four-fold degenerate in the effective-mass approximation. Due to valley-orbit interaction the ground state

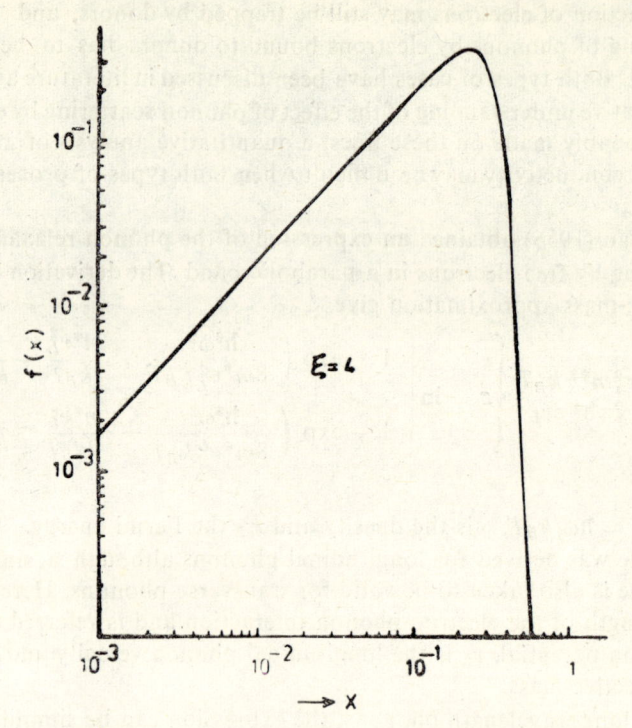

Fig. 5.6 The dependence of phonon-electron scattering on reduced phonon frequency $x = \omega/\omega_D$. The function $f(x)$ is given by

$$f(x) = \frac{\hbar \omega_D x}{k_B T} - \ln\left[\frac{1 + \exp\left(\psi + \frac{\hbar \omega_D x}{2k_B T}\right)}{1 + \exp\left(\psi - \frac{\hbar \omega_D x}{2k_B T}\right)}\right]$$

where $\psi = \dfrac{\hbar^2 \omega^2}{8m^* v_L^2 k_B T} + \dfrac{m^* v_L^2}{2k_B T} - \xi$ (after Parrott, 1979).

is split into a singlet and a triplet separated by 4Δ (the chemical shift). The energy change of the electrons is substantial in the presence of the strain.

The energy due to a static strain-field is proportional to the square of the strain. Keyes obtained for the relaxation time

$$\tau^{-1} \propto \omega^4 (1 + q^2 r_0^2/4)^{-8} \tag{5.26}$$

in the acoustic approximation: $\omega = v_s q$, r_0 is the mean radius of the localized state. When q increases to more than $1/r_0$, the scattering decreases very quickly. This cut-off could explain the steepness of the slope of the lattice thermal conductivity versus T curve below the conductivity maximum.

Griffin and Carruthers (1963) proposed a more refined version of the theory in which the dynamic response of the electron is considered in a rigorous way. The electron-phonon interaction induces virtual transitions of the bound electron to higher excited states. The fact that 4Δ lies within the range of heat conduction clearly shows that the resonance would be significant. They obtained for the relaxation time

$$\tau_{pe}^{-1} = \text{const } \varepsilon_4^2 \, n_{ex} \, (4\Delta)^2 \frac{\omega^4}{(1 + r_0 \, \omega^2/4v_s^2)^8} \frac{1}{(\hbar^2\omega^2 - (4\Delta)^2)^2} \qquad (5.27)$$

ε_4 is the shear deformation potential and n_{ex} is the concentration of uncompensated donor electrons.

This expression differs from the one derived by Keyes mainly by the resonance term.

The large thermal resistance, which arises from the scattering of phonons by electrons, has been observed in n-type Ge and GaSb at low temperatures (Goff and Pearlman, 1965, Mathur and Pearlman, 1969, and Poujade and Albany, 1969).

Kwok (1966) studied acoustic attenuation by neutral donors in germanium and pointed out that the elastic scattering of phonons by electrons in donor levels (as calculated on the basis of the theory given by Keyes and Griffin and Carruthers) was too small to account for the attenuation of transverse acoustic phonons at the microwave frequencies in slightly doped n-type germanium (Pomeranz, 1965). He calculated the acoustic attenuation due to inelastic phonon scattering and thermally assisted absorption, and found good agreement with the experimental data. His expression for the relaxation time due to neutral donor scattering includes both elastic and inelastic scattering from the ground state and the next excited state. Resonance absorption of phonons was also investigated by Pomeranz using the Green's function technique.

5.10.4 *Further considerations*

Ziman's expression for the phonon-electron scattering relaxation time can be expressed in the form

$$\tau_{pe}^{-1} = \Omega(q) \, F(q) \qquad (5.28)$$

$F(q)$ is a cutoff function which takes account of the fact that not every phonon can be scattered by electrons Consider a situation in which an electron in state **k** is scattered to a state **k'**, by absorbing a phonon

$$\mathbf{k'} = \mathbf{k} + \mathbf{q}$$

Both **k** and **k'** can take the upper limiting value k_F and this sets a limit $q \leqslant 2k_F$ on **q**. This may yield a cutoff function $F(q) = 1$ for $q \leqslant 2k_F$ and $F(q) = 0$ for $q > 2k_F$.

The scattering function $\Omega(q)$ takes the form

$$\Omega(q) = Aq \qquad (5.29)$$

where $A = m^{*2}\varepsilon_1^2/2\pi\rho\hbar^3$.

This expression taken along with other phonon-scattering processes gives a relatively good fit to the observed thermal conductivity at the highest temperatures. At low temperature the simple relationship $\Omega(q) = Aq$ overestimates the scattering of low q phonons. When the phonon wavelength is of the order of or greater than the carrier mean-free-path a decrease in the strength of electron-phonon interaction is expected. The interaction may also be affected by the screening due to the carriers.

Following the suggestion by Crosby and Grenier (1971), the screening of the hole-phonon interaction for the low-frequency phonons in heavily-doped p-type InSb was discussed by Fozooni et al (1980). They have given a modified version of Ziman's expression for the phonon-electron scattering relaxation time as

$$\tau_{pe}^{-1} = \frac{B(ql) \, Aq \, F(q)}{[1 + 3(q_{TF}/q)^2]^2} \qquad (5.30)$$

Here l is the carrier mean-free-path, $q_{TF} = \omega_p/v_F$ the Thomas-Fermi wavevector, ω_p the plasma frequency and v_F is the Fermi velocity. This differs from the earlier expression in the two terms: $[1 + 3(q_{TF}/q)^2]^{-2}$ and $B(ql)$. The first term is the screening factor and equal to unity for $q \gg q_{TF}$, but falls off as $(q/q_{TF})^4$ for $q < q_{TF}$.

The second term $B(ql)$ is given by

$$B(ql) = \frac{2ql}{\pi} \frac{\tan^{-1} ql}{ql - \tan^{-1} ql} \qquad (5.31)$$

For large ql, the term $B(ql) \simeq 1$, and for small ql it increases as $(ql)^{-1}$ and tends to limit the screening.

Kosarev (1971) takes into account the electric field of the ionized impurities, in the presence of which the phonon-carrier interactions take place. This allows the Ziman cut-off condition $q < 2k_F$ to be relaxed and the high frequency phonons to interact with the carriers. The thermal conductivity data in heavily doped p-type GaAs and InSb have been successfully analysed on these lines (Singh and Verma 1974, Fozooni et al, 1980).

Sota and Suzuki (1982, 1983) derived an expression for τ_{pe} in heavily doped many valley semiconductors and applied their theory to n-type germanium and silicon at 0 K. This theory is expected to describe the phonon-electron scattering both in the high as well as low-frequency regions. The theory was extended (Sota and Suzuki, 1984) to p-type material which takes into account the interband hole-phonon interaction, the intra-band hole-phonon interaction and both the deformation potential and the piezoelectric coupling. The shear components of the hole-phonon interaction are shown to play an important role in small wave-number region.

References

Abeles, B. (1963), *Phys. Rev. 131*, 1906.
Anderson, A.C. and Malinowski, M.E. (1972), *Phys. Rev. B5*, 3199.

Bak, Thor. A. (ed) (1964), Phonons and Phonon Interactions, *Aarhus Summer School Lectures*, 1963, W.A. Benjamin, Inc., New York.
Bell, R.J. (1972), *Rep. Progr. Phys. 35*, 1315.
Berman, R. and Brock, J.C.F. (1965), *Proc. Roy. Soc. A289*, 46.
Bottger, H. (1974), *Phys. Stat. Solidi (b) 62*, 9.
Brockhouse, B.N. (1964), Phonons and Phonon Interactions, *Aarhus Summer School Lectures*, 1963, W.A. Benjamin, Inc., New York.
Brockhouse, B.N. (1966), *Phonons in Perfect Lattices and Lattices with Point Imperfections* (ed. R.W.H. Stevenson), Scottish universities Summer School, 1965, Oliver and Boyd, London.
Bross, H., Seeger, A. and Haberkorn, R. (1963), *Phys. Stat. Solidi 3*, 1126.
Callaway, J. (1959), *Phys. Rev. 113*, 1046.
Callaway, J. (1963), *Nuovo Cim. 29*, 883.
Carruthers, P. (1961), *Rev. Mod. Phys. 33*, 92.
Cochran, W. (1959), *Proc. Roy. Soc. A253*, 260.
Crosby, C.R. and Grenier, C.G. (1971), *Phys. Rev. B4*, 1258.
Dean, P. (1972), *Rev. Mod. Phys. 44*, 127.
Dolling, G. (1963), *Int. Atomic Energy Agency*, Vienna 1, 37.
Ecsedy, D.J. and Klemens, P.G. (1975), *Bull. Am. Phys. Soc. 20*, 356.
Fozooni, P., Zebouni, N.H. and Grenier, C.G. (1980), *J. Phys. C: Solid St. Phys. 13*, 4285.
Goff, J.F. and Pearlman, N. (1965), *Phys. Rev. 140*, A 2151.
Griffin, G. and Carruthers, P. (1963), *Phys. Rev. 131*, 1976.
Herpin, A. (1952), *Ann. Phys. 7*, 97.
Herring, C. (1954), *Phys. Rev. 95*, 954.
Holland, M.G. (1963), *Phys. Rev. 132*, 2461.
Holland, M.G. and Neuringer, L.J. (1962), *Proc. of Internationol Conf. on the Physics of Semiconductors*, Exeter, p 35, Inst. of Physics and Physical Society, London.
Hori, J. (1968), *Spectral Properties of Disordered Chains and Lattices*, Pergamon Press, Oxford.
Ishioka, S. and Suzuki, H. (1963), *J. Phys. Soc. Japan 18*. (suppl. II), 93.
Joshi, Y.P., Tewari, M.D. and Verma, G.S. (1970), *Phys. Rev. B1*, 642.
Keyes, R.W. (1961), *Phys. Rev. 122*, 1171.
Keyes, R.W. (1962), *J. Appl. Phys. 33*, 3371.
Klein, M.V. (1963), *Phys. Rev. 131*, 1500.
Klein, M.V. (1966), *Phys. Rev. 141*, 716.
Klemens, P.G. (1951), *Proc. Roy. Soc.* (London) A208, 108.
Klemens, P.G. (1955), *Proc. Phys. Soc.* (London) A68, 1113.
Klemens, P.G. (1958), *Solid St. Phys.* (ed. F. Seitz and D. Turnbull) Academie Press, vol 7, 1.
Klemens, P.G. and Ecsedy, D.J. (1976), *Phonon Scatt. in Solids* (eds. L.J. Challis, V.W. Rampton and A.F.G. Wyatt), Plenum Press, New York.
Krumhansl, J.A. (1963), *Proc. of Int. Conf. on Lattice Dynamics*, Copenhagen (ed. R.F. Wallis), Pergamon Press, London, p. 523.
Kosarev, V.V. (1971), *Sov. Phys.—JETP 33*, 793.
Kwok, P.C. (1966), *Phys. Rev. 149*, 666.
Lifshitz, I.M. (1944), *J. Phys. USSR 8*, 89.

Maradudin, A.A. (1966), *Solid St. Phys.* (eds. F. Seitz and D. Turnbull), Vol. 18, 273 and Vol. 19, 1.

Maradudin, A.A., Montroll, E.W. and Weiss, G.H. (1963), *Theory of Lattice Dynamics in the Harmonic Approximation, Solid St. Phys.* (ed. F. Seitz and D. Turnbull), Suppl. 3.

Mathur, M.P. and Pearlman, N. (1969), *Phys. Rev. 180*, 833.

Mozer, B., Otnes, K. and Myers, V.W. (1962), *Phys. Rev. Lett. 8*, 278.

Nabarro, F.R.N. (1951), *Proc. Roy. Soc. A209*, 278.

Ohashi, K. (1968), *J. Phys. Soc. Japan 24*, 437.

Parrott, J.E. (1979), *Rev. int. hautes temper. Refract. Fr. 16*, 393.

Pohl, R.O. (1962), *Phys. Rev. Lett. 8*, 48.

Pomeranchuk, I. (1941), *Phys. Rev. 60*, 820.

Pomeranchuk, I. (1942), *J. Phys. USSR 5*, 237.

Pomerantz, M. (1965), *Proc. IEEE 53*, 1438.

Poujade, A.M. and Albany, J.H. (1969), *Phys. Rev. 182*, 802.

Schaefer, G. (1960), *J. Phys. Chem. Solids 12*, 233.

Singh, M. and Verma, G.S. (1974), *J. Phys. C: Solid St. Phys. 7*, 3743.

Sota, T. and Suzuki, K. (1982), *J. Phys. C: Solid St. Phys. 15*, 6991.

Sota, T. and Suzuki, K. (1983), *J. Phys. C: Solid St. Phys. 16*, 4347.

Sota, T. and Suzuki, K. (1984), *J. Phys. C: Solid St. Phys. 17*, 2661.

Sproull, R., Moss, M. and Weinstock, H. (1959), *J. Appl. Phys. 30*, 334.

Steigmeier, E.F. and Kudman, I. (1966), *Phys. Rev. 141*, 767.

Stevenson, R.W.H. (1966) (ed.), *Phonons in Perfect Lattices and in Lattices with Point Imperfections*, Scottish Univ. Summer School, 1965, Oliver and Boyd, London.

Suzuki, T. and Suzuki, H. (1972), *J. Phys. Soc. Japan 32*, 164.

Takeno, S. (1963), *Progr. Theor. Phys. 29*, 191.

Vook, F.L. (1965), *Phys. Rev. 140*, A2013.

Wagner, M. (1963), *Phys. Rev. 131*, 1443.

Wagner, M. (1964), *Phys. Rev. 133*, A750.

Walker, C.T. and Pohl, R.O. (1963), *Phys. Rev. 131*, 1433.

Warren, J.L., Wenzel, R.G. and Yarnell, J.L. (1965), *Inelastic Scattering of Neutrons*, I.A.E.A., Vienna.

Yussouf, M. and Mahanty, J. (1965), *Proc. Phys. Soc. 85*, 1223.

Yussouf, M. and Mahanty, J. (1966), *Proc. Phys. Soc. 87*, 689.

Ziman, J.M. (1956), *Phil. Mag. 1*, 191; *Phil. Mag. 2*, 292 (1957).

Ziman, J.M. (1960), *Electrons and Phonons*, Clarendon Press, Oxford.

Chapter VI

Lattice Thermal Conductivity

6.1 Introduction

Early theories of thermal conductivity were presented by Debye (1914) and Peierls (1929); other pioneering papers in this field were published by Akhiezer (1940), Pomeranchuk (1941), Klemens (1951, 1955 and 1960), Leibfried and Schlomann (1954), Dugdale and MacDonald (1955), Callaway (1959), Sussman and Thelung (1963), Erdos (1965), Holland (1963), Ranninger (1965) and Julian (1965). Several review papers by Klemens (1958), Carruthers (1961), Bross (1962), Holland (1966), Klemens (1969), Steigmeier (1969) and Slack (1979) have dealt with the various aspects of the subject. Texts by Ziman (1960), Parrott and Stuckes (1975) and Berman (1976) describe the theory of thermal conductivity in varying detail.

The theoretical methods used in calculating lattice thermal conductivity are described in this chapter. The salient features of thermal conductivity of perfect nonmetallic crystals are discussed within the framework of the variational approach. The relaxation time method is also described and generalized expressions obtained for the lattice thermal conductivity. The applicability of this method in analysing the experimental data of a variety of crystalline semiconductors forms the subject matter of the next two chapters.

6.2 The variational method

6.2.1 *Introduction*

The essential features in obtaining a solution of the Boltzmann equation* for phonons are similar to that for electrons (described in Chap. III). The thermodynamic aspects as applied to the flow of charge and energy have

---

* A more direct approach to the calculation of thermal conductivity is based on the Kubo method (Schieve and Peterson 1962, and Maradudin, 1964). However, this approach has not proved to be very successful and, in spite of their intuitive nature, the methods based on the Boltzmann equation are widely used in the calculation of thermal conductivity.

been described for the electron system and the basic methodology is the same for the phonon system. The internal production rate of entropy by the scattering processes has to be obtained and equated to the rate of entropy production due to the heat flowing down the temperature gradient.

Leibfried and Schlomann (1954) and Ziman (1956 and 1960) were the first to apply this method to the lattice conduction although it had earlier been used to calculate the electrical conductivity. Hamilton and Parrott (1969) applied this method to calculate the thermal conductivity of germanium. Srivastava and Hamilton (1978) have reviewed the complementary variational principles with application to the theory of thermal transport.

In the relaxation-time method, the expression for the lattice thermal conductivity (Eq. 6.39) does not have the nature of the phonon distribution explicitly contained in it although the relaxation time $\tau(\mathbf{q})$ depends on such a distribution. In the variational method, the nature of the phonon distribution is of primary importance.

The phonon distribution is displaced from its equilibrium value ($N_\mathbf{q}^0$) due to various causes. A function $\phi_\mathbf{q}$ is introduced which measures the deviation from the equilibrium distribution.

$$N_\mathbf{q} = N_\mathbf{q}^0 - \phi_\mathbf{q} \frac{\partial N_\mathbf{q}^0}{\partial E_\mathbf{q}} \tag{6.1}$$

Differentiating Eq. 5.6 with respect to $E_\mathbf{q}$ one obtains $\partial N_\mathbf{q}^0/\partial E_\mathbf{q}$ which, when substituted in Eq. 6.1, gives

$$N_\mathbf{q} = N_\mathbf{q}^0 + \phi_\mathbf{q} \frac{N_\mathbf{q}^0 (1 + N_\mathbf{q}^0)}{k_B T} \tag{6.2}$$

6.2.2 *Elastic scattering from state* q *to* q′

We shall now try to obtain the Boltzmann equation for the case when a phonon in the state $\mathbf{q}$ is scattered to the state $\mathbf{q}'$. If $\mathbf{q}'$ lies in the range $d\mathbf{q}'$, the transition probability can be written as

$$P_\mathbf{q}^{\mathbf{q}'} d\mathbf{q}' = N_\mathbf{q} (1 + N_{\mathbf{q}'}) Q_\mathbf{q}^{\mathbf{q}'} d\mathbf{q}' \tag{6.3}$$

Apart from the intrinsic transition rate $Q_\mathbf{q}^{\mathbf{q}'}$ the population factor $N_\mathbf{q}(1 + N_{\mathbf{q}'})$ has to be taken into consideration (Ziman, 1960). The rate of change of the distribution function is then given by

$$\left(\frac{\partial N_\mathbf{q}}{\partial t}\right)_\text{scatt} = \int \left[N_{\mathbf{q}'} (1 + N_\mathbf{q}) Q_{\mathbf{q}'}^{\mathbf{q}} - N_\mathbf{q} (1 + N_{\mathbf{q}'}) Q_\mathbf{q}^{\mathbf{q}'} \right] d\mathbf{q}' \tag{6.4}$$

We sum over all states $\mathbf{q}'$ from which a phonon may come to the state $\mathbf{q}$ and into which it may go. Using the principle of microscopic reversibility, $Q_\mathbf{q}^{\mathbf{q}'} = Q_{\mathbf{q}'}^{\mathbf{q}}$, and since the phonon energy remains unchanged in an elastic scattering process leading to $N_\mathbf{q}^0 = N_{\mathbf{q}'}^0$, we can write after some simplification

$$\left(\frac{\partial N_q}{\partial t}\right)_{scatt} = \frac{1}{k_B T}\int (\phi_{q'} - \phi_q)\, P_q^{q'}\, dq' \tag{6.5}$$

The corresponding Boltzmann equation is then written as

$$-v_q \cdot \nabla T \frac{\partial N_q^0}{\partial T} = \frac{1}{k_B T}\int (\phi_{q'} - \phi_q)\, P_q^{q'}\, dq' \tag{6.6}$$

6.2.3 Three-phonon processes
For a three-phonon process described by

$$(\mathbf{q}, s) + (\mathbf{q}', s') \rightleftharpoons (\mathbf{q}'', s'') \tag{6.7}$$

the transition rate is given by $N_q N_{q'}(1 + N_{q''})\, Q_{q,q'}^{q''}$.

The Boltzmann equation then takes the form

$$-v_q \cdot \nabla T \frac{\partial N_q}{\partial T} = \iint [\{N_q N_{q'}(1 + N_{q''}) - (1 + N_q)(1 + N_{q'}) N_{q''}\} Q_{q,q'}^{q''}$$
$$+ \tfrac{1}{2}\{N_q(1 + N_{q'})(1 + N_{q''}) - (1 + N_q) N_{q'} N_{q''}\} Q_q^{q',q''}]\, dq'\, dq'' \tag{6.8}$$

This includes both types of processes: one in which the phonon $(\mathbf{q}, s)$ is split into two phonons and the other when two phonons combine to create a phonon.

6.2.4 Umklapp processes
It has been well understood that the three-phonon umklapp processes form the most significant contribution to the phonon scattering in limiting high temperature thermal conductivity.

The basic three-phonon process has been described in the previous section. Variational integrals can be written for this process and the rate of entropy production obtained for the internal processes together with that due to the flow of heat down the temperature gradient. A comparison of the two expressions for the entropy production (Ziman, 1960) gives

$$\frac{1}{\lambda} = \frac{1}{2k_B T^2} \frac{\iiint \{\phi_q + \phi_{q'} - \phi_{q''}\}^2 P_{q,q'}^{q''}\, dq\, dq'\, dq''}{\left|\iint v_q \phi_q \frac{\partial N_q^0}{\partial T}\, dq\right|^2} \tag{6.9}$$

This must be a minimum when ϕ_q satisfies the Boltzmann equation.

N-process
For the momentum conserving processes, the trial function

$$\phi_q = \mathbf{q} \cdot \mathbf{u} \tag{6.10}$$

where $\mathbf{u}$ is a unit vector along ∇T, gives

$$\{\phi_q + \phi_{q'} - \phi_{q''}\} = (\mathbf{q} + \mathbf{q}' - \mathbf{q}'') \cdot \mathbf{u} \tag{6.11}$$

and this vanishes for all N-processes. This choice of ϕ_q does not give a zero value for the denominator in Eq. 6.9 which would result in a zero thermal resistivity.

6.2.5 Umklapp resistance

The trial function for this can be written on similar lines. The term

$$\{\phi_q + \phi_{q'} - \phi_{q''}\} = (q + q' - q'') \cdot u$$

would take the value $\mathbf{G} \cdot \mathbf{u}$ (where $\mathbf{G}$ is the reciprocal lattice vector) and would thus naturally exclude the terms relating to N-processes.

The denominator of Eq. 6.9 is the square of the modulus of

$$\frac{\partial}{\partial T}\left\{\int v_q (\mathbf{q} \cdot \mathbf{u}) N_q^0 \, dq\right\} = \frac{1}{3\hbar}\frac{\partial}{\partial T}\left\{\int E_q N_q^0 \, dq\right\} = C/3\hbar \qquad (6.12)$$

It requires some special effort to deal with the integral in the numerator. The integration over $\mathbf{q}'$ and $\mathbf{q}''$ can be reduced to a double integral over the energy conservation surface S'—the locus of the allowed values of $\mathbf{q}'$. The rate of entropy production is written as (Ziman, 1960)

$$\dot{S}_{\text{scatt}} = \frac{1}{2k_B^2 T}(\overline{\mathbf{G} \cdot \mathbf{u}})^2 \frac{V}{4\pi^2 \hbar^2} \iint N_q^0 N_{q'}^0 (1$$

$$+ N_{q''}^0) \mid F_{q,p; q',p'; q'',p''} \mid^2 \frac{dS'}{v_n'} \, dq \qquad (6.13)$$

v_n' is the sound velocity in a direction normal to S'.

The function F appears in the expression for the transition probability and is a measure of the strength of the phonon-phonon interaction. Assuming a continuum model, F can be written in terms of the anharmonicity tensor. One then takes all components of this tensor into a single average anharmonicity factor. All sound velocities are gathered together and related to the density ρ_0. One then obtains after some rearrangement (Ziman, 1960)

$$W_U \sim \frac{G^2 \hbar^3 \gamma^2}{k_B T^2 \rho_0 C^2} \iint N_q^0 N_{q'}^0 (1 + N_{q''}^0) \, q \, q' \, q'' \, dS' \, dq \qquad (6.14)$$

The anharmonicity of the lattice potential has been related to the Gruneisen parameter γ.

The equilibrium phonon distribution can at high temperatures be written as

$$N_q^0 \sim k_B T / \hbar v_s q \qquad (6.15)$$

If N is the number of atoms per unit volume the heat capacity is given by $C = 3Nk_B$. With these simplifications the umklapp resistivity can at high temperature be written as

$$W_U \sim \frac{G^2 \hbar^3 \gamma^2}{k_B T^2 \rho_0 N^2 k_B^2}\left(\frac{k_B T}{\hbar v_s}\right)^3 \iint dS' dq$$

$$\sim \frac{G^2 \gamma^2 T}{\rho_0 N^2 v_s^3} \iint dS' dq \qquad (6.16)$$

In order to evaluate the integral for each value of **q** one has to determine the area of the energy conservation surface and sum over all **q**. The result is $\sim G^2 \times G^3$, since **q** goes over the volume of a zone and S' must be an appreciable fraction of the area of a face of the zone. Further $G \sim 1/a$ where a is the lattice constant and $G^3 \sim N$, one obtains

$$\lambda \sim \frac{\rho_0 v_s^3 a}{\gamma^2 T} \qquad (T > \theta_D) \tag{6.17}$$

Leibfried and Schlomann (1954) obtained a similar expression for λ*. Expressing the sound velocity v_s in terms of the characteristic Debye temperature θ_D, a slightly different expression can be obtained for λ

$$\lambda = \lambda_0 f(\theta_D/T) \tag{6.18}$$

where $\lambda_0 = B\overline{M}\delta\theta_D^2\gamma^{-2}$ is the thermal conductivity at the Debye temperature, B is a constant, $\overline{M}$ is the mean atomic weight, and δ is related to the lattice constant a.

At low temperatures the lattice thermal conductivity is given by

$$\lambda \simeq (T/\theta_D)^3 \exp(\theta_D/bT) \qquad (T < \theta_D) \tag{6.19}$$

b is a numerical parameter. Combining the high and the low temperature results (i.e. Eq. 6.18 and 6.19) we can write

$$\lambda = \lambda_0 f(\theta_D/T) \tag{6.20}$$

where
$$f(\theta_D/T) = \theta_D/T, \quad \text{for} \quad T > \theta_D$$
$$= (T/\theta_D)^3 \exp(\theta_D/bT) \quad \text{for } T < \theta_D$$

6.2.6 Elastic scattering by defects

The Boltzmann equation for this case has already been described in subsection 6.2.2. The rate of entropy production is given by (Ziman, 1960 and Berman, 1976)

$$\left(\frac{\partial S}{\partial t}\right)_{scatt} = \frac{1}{2k_B T^2} \iint \{\phi_\mathbf{q} - \phi_{\mathbf{q}'}\}^2 P_\mathbf{q}^{\mathbf{q}'} d\mathbf{q}\, d\mathbf{q}' \tag{6.21}$$

The thermal conductivity is obtained by equating the two rates of the entropy production and is given by

$$\frac{1}{\lambda} = \frac{\frac{1}{2k_B T^2} \iint \{\phi_\mathbf{q} - \phi_{\mathbf{q}'}\}^2 P_\mathbf{q}^{\mathbf{q}'} d\mathbf{q}\, d\mathbf{q}'}{\left[\frac{1}{T} \int E_\mathbf{q} v_\mathbf{q} \phi_\mathbf{q} \frac{\partial N_\mathbf{q}^0}{\partial E_\mathbf{q}} d\mathbf{q}\right]^2} \tag{6.22}$$

For this type of scattering a suitable trial function is

---

*As this chapter is almost exclusively devoted to the lattice thermal conductivity, λ_L will be written as λ to avoid multiplicity of subscripts.

$$\phi_q = \frac{1}{q^n}(\mathbf{q}\cdot\mathbf{u}) \tag{6.23}$$

There are several forms of ϕ_q which give the same minimum value, $1/\lambda = 0$. For $n \geqslant 3.5$ both the numerator and the denominator of Eq. 6.22 diverge as $q \to 0$ but maintain a ratio equal to zero. If we assume that the distribution does not extend to $q = 0$, but stops at $q = q_{min}$, one obtains

$$1/\lambda \propto q_{min}\frac{(n-3)^2}{(2n-7)} \tag{6.24}$$

For a positive resistance $n \geqslant 3.5$, and the minimum of $1/\lambda$ occurs for $n = 4$. If q_{min} is now lowered indefinitely, $1/\lambda$ is still a minimum for $n = 4$ and this minimum is zero for $q_{min} = 0$.

This is certainly the minimum value which we can hope to find and this will therefore be the thermal resistivity if the only scattering mechanism present is the defect scattering. The function ϕ_q for this case is then given by $(\mathbf{q}\cdot\mathbf{u})/q^4$.

N-processes and point-defect scattering
So far situations with only one type of scattering mechanism have been discussed. In situations when two or more scattering mechanisms are present the sum of appropriate integrals are included in the numerator of the variational expression.

When three-phonon normal processes are considered along with the defect scattering, the thermal conductivity expression is written as

$$\frac{1}{\lambda} = \frac{\frac{1}{2k_BT^2}\left[\iint\{\phi_q-\phi_{q'}\}^2 P_q^{q'}\,dq\,dq' + \iiint\{\phi_q+\phi_{q'}-\phi_{q''}\}^2 P_{q,q'}^{q''}\,dq\,dq'\,dq''\right]}{\left[\frac{1}{T}\int E_q\,v_q\,\phi_q\frac{\partial N_0}{\partial E_q}\right]^2} \tag{6.25}$$

Earlier discussions described the trial functions for the two cases and in either case a zero value of resistivity was obtained. With either of these choices one of the terms would acquire a nonzero value. A suitable combination of the two was proposed by Berman et al. (1959). Above a certain value of $q = q_0$, ϕ_q is governed by both the processes as if they were acting separately. However, as the defect case results in a large deviation from equilibrium for low $\mathbf{q}$ values, it is safe to assume that at q_0 the contribution to ϕ_q reaches a limiting value. In that case

$$\phi_q = a_0(\mathbf{q}\cdot\mathbf{u}) + \frac{a_4}{q^4}(\mathbf{q}\cdot\mathbf{u}) \quad \text{for } q \geqslant q_0$$

$$\phi_q = a_0(\mathbf{q}\cdot\mathbf{u}) + \frac{a_4}{q_0^4}(\mathbf{q}_0\cdot\mathbf{u}) \quad \text{for } q < q_0 \tag{6.26}$$

6.3 Absolute magnitudes

6.3.1 *Thermal conductivity of perfect non-metallic crystals*

It is of interest to make a comparison between the calculated and observed values of thermal conductivities for perfect non-metallic crystals. Slack (1979) has reviewed various aspects of the thermal conductivity behaviour for a large number of crystals.

Referring to Eq. 6.18 the lattice thermal conductivity λ at $T = \theta_D$ can be written as

$$\lambda(\theta_D) = B\bar{M} \delta \theta_D^2 \gamma^{-2} \qquad (6.27)$$

where θ_D and γ values appropriate for the acoustic mode were taken into consideration. The constant B has a value 5.7×10^{-8} (Leibfried and Schlomann, 1954) when λ is to be in CGS units and δ in angstrom units. In MKS units with δ expressed in nanometres, B has a value 5.7×10^{-5}.

Julian (1965) suggested a slightly different approach in which θ_D is replaced by $\theta_{D\infty}$, which, for rare gas crystals, forms the high temperature limit of θ_D for the acoustic modes. Using $\theta_{D\infty}$ obtained from the phonon density-of-states, the thermal conductivity can be written as

$$\lambda(\theta_{D\infty}) = B\bar{M} \delta\theta_{D\infty}^2 \gamma^{-2} \qquad (6.28)$$

B has a value 3×10^{-5} with λ in watt per metre per Kelvin and δ in nanometres.

Theoretical values of the thermal conductivity obtained for the rare gas crystals (with one atom per unit cell) compare well (within $\pm$ 20%) with the experimental data.

Slack (1979) extended this type of comparison to more complex crystals with $n = 2, 3$, etc. where n is the number of atoms per primitive cell. For $n = 1$, only acoustic phonon modes have to be considered, whereas for $n \geqslant 2$, optic phonon branches also need to be taken into consideration. Neutron scattering measurements provide detailed information about the various phonon branches. In a large number of crystals, such as Si, Ge, GaAs, GaSb, etc. the group velocity associated with optical phonons was found to be much smaller than that of the acoustic phonons. It was, therefore, reasonable to assume that all heat is carried by acoustic phonons although in some materials like AlSb, about 15 per cent of the total heat may be carried by optic phonons (Wagini, 1966). However, the contribution of optic phonons has been neglected in most of the theoretical analysis. Some aspects of this topic are described in the appendix.

Assuming that heat is carried only by acoustic phonons and that they interact only with other acoustic phonons, a simple generalization of Eq. 6.28 can be obtained for $n \geqslant 2$. To find and include all of the acoustic phonons and their interactions, a counting procedure was adopted (Slack, 1972, and Roufosse and Klemens, 1973). The acoustic branch thermal conductivity of an fcc crystal is then obtained as (Slack, 1979)

$$\lambda(\tilde{\theta}_{D\infty}) = Bn^{1/3} \bar{M} \delta \tilde{\theta}_{D\infty}^2 \tilde{\gamma}^{-2} \qquad (6.29)$$

106 THERMAL CONDUCTION IN SEMICONDUCTORS

The value of the Debye temperature appropriate to the acoustic phonons ($\tilde{\theta}_{D\infty}$) has to be used and similarly γ must refer to the acoustic branches only at high temperatures ($\tilde{\gamma}_\infty$). The Debye temperature at $0\,K$ ($\tilde{\theta}_{D_0}$) is related to the standard value (θ_{D_0}) by

$$\tilde{\theta}_{D_0} = n^{-1/3}\,\theta_{D_0} \qquad (6.30)$$

Figure 6.1 shows a comparison between the calculated and the experimental values of λ for cubic crystals having two atoms per unit cell at $T = \tilde{\theta}_{D\infty}$, which is the high temperature value of the Debye temperature corresponding to acoustic phonons. A large number of semiconductors fall into this category and, except in a few cases, the agreement between theoretical and experimental values is not much worse than that for the $n = 1$ case.

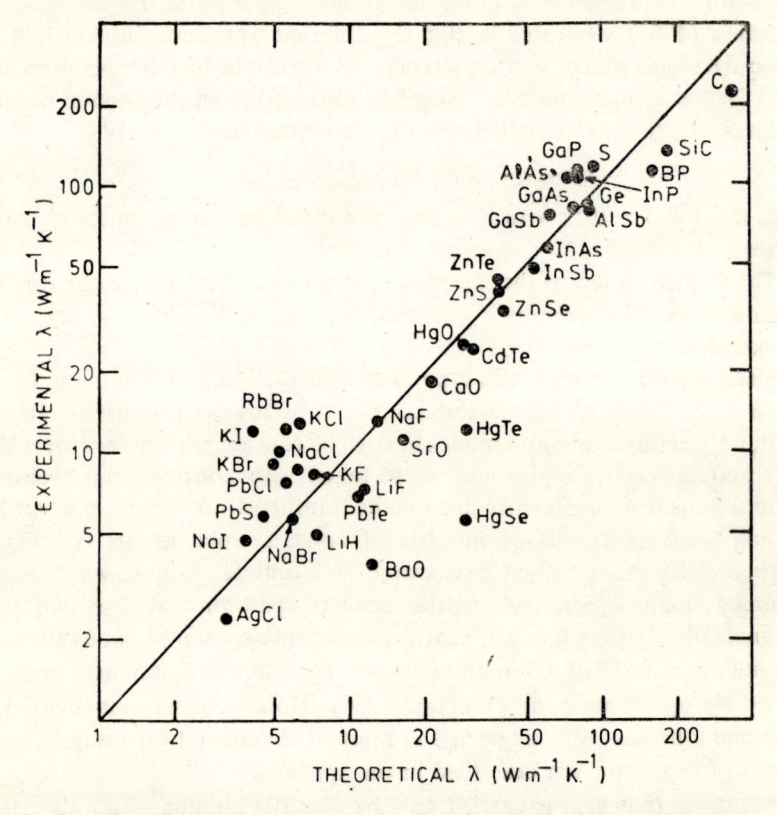

Fig. 6.1 Experimental and theoretical values for the thermal conductivity of crystals of adamantine and rocksalt structures at $T = \tilde{\theta}_{D\infty}$ (after Slack, 1979).

Another generalization of the thermal conductivity expression is given by (Slack, 1979)

$$\lambda(T) = B\overline{M}\,\delta\,\theta_{D0}^3\,n^{-2/3}\,T^{-1}\,\gamma^{-2} \qquad (6.31)$$

The values of θ_{D_0} and γ depend on $\bar{M}$, δ and the interatomic forces, and not on n. Hence, for similar crystals, λ decreases with an increase in n at a fixed temperature. Oliver and Slack (1966) earlier introduced a crystal complexity factor to account for such a decrease in λ as the crystal structure becomes more complex.

6.3.2 Transverse and longitudinal phonons

In the earlier attempts to explain the measured thermal conductivity no distinction was made between different phonon polarizations. With a more detailed knowledge of the phonon dispersion curves, it became possible to separate the contributions to heat transport from the longitudinal and transverse branches of the phonon spectrum.

To take into consideration the separate contributions from different polarization branches the parameters, θ_D and γ should be known corresponding to the branches. If γ is taken to be the same for the two polarizations it is easy to separate the contributions to the density-of-states $g(\nu)\,d\nu$. The Debye temperature is now given by

$$(\tilde{\theta}_{D\infty}^{T,L})^2 = \frac{5h^2}{3k_B^2} \frac{\int \nu^2 g_{T,L}(\nu)\,d\nu}{\int g_{T,L}(\nu)\,d\nu} \tag{6.32}$$

$g_{T,L}(\nu)$ is the density-of-states for transverse and longitudinal phonons. This leads to

$$(\tilde{\theta}_{D\infty})^2 = \tfrac{1}{3}[2(\tilde{\theta}_{D\infty}^T)^2 + (\tilde{\theta}_{D\infty}^L)^2]$$

With these relationships and assuming that the Gruneisen parameter has the same value for the two polarizations, it can be shown that

$$\frac{\lambda_L}{\lambda_T} = \frac{1}{2}\left(\frac{\tilde{\theta}_{D\infty}^L}{\tilde{\theta}_{D\infty}^T}\right)^2 \tag{6.33}$$

For GaAs the ratio is about 3.0 and for KCl it has a value of about 1.0. The conclusion reached on the basis of these arguments is that the two phonon branches contribute about equally to the total thermal conductivity. The qualitative results appear to be reasonable but an accurate value of λ_L/λ_T would require information about the separate scattering rates for the two polarizations. This aspect of the problem will be taken up again in Sec. 6.6.

6.4 Lattice thermal conductivity—the relaxation-time method

Having assumed the existence of a well-defined relaxation time $\tau(\mathbf{q})$ (see Chap V, Sec. 5.6), obtaining a generalized expression for the lattice thermal conductivity becomes a straightforward operation.

Assuming that the distribution N_q does not deviate too far from its equilibrium value N_q^0, it is convenient to replace $\partial N_q/\partial T$ by $\partial N_q^0/\partial T$. This gives

$$\left(\frac{\partial N_q^0}{\partial t}\right)_{\text{drift}} = - v_x \frac{\partial N_q^0}{\partial T} \frac{\partial T}{\partial x} \qquad (6.34)$$

The heat flux is then given by

$$W = - \sum_{q,s} \hbar \omega_s(q) \, (v_x)^2 \tau_s(q) \frac{\partial N_{qs}^0}{\partial T} \frac{\partial T}{\partial x} \qquad (6.35)$$

The polarisation index s has now been reintroduced. The lattice thermal conductivity can be obtained by dividing W by the negative temperature gradient and gives

$$\lambda = \sum_{q,s} \hbar \omega_s(q) \, (v_x)^2 \tau_s(q) \frac{\partial N_{qs}^0}{\partial T} \qquad (6.36)$$

It is convenient to replace the summation over q by an integral over ω. If $g(\omega) d\omega$ is the number of phonon modes in the frequency interval ω and $\omega + d\omega$ and with the substitution, $v_x^2 = v_s^2/3$, we get

$$\lambda = \tfrac{1}{3} \sum_s \int \hbar \omega v_s^2 \, \tau_s(\omega) \, g_s(\omega) \frac{\partial N_0(\omega)}{\partial T} d\omega \qquad (6.37)$$

This equation is based on the assumption that the substance is isotropic. Detailed information on the frequency distribution is available for various solids and by incorporating this information into Eq. 6.37 the lattice thermal conductivity can be calculated rigorously. However, a simple generalized expression is useful and often justifies the somewhat crude assumptions for its simplification. The Debye model provided such a simplification by assuming a linear dispersion relation and taking the density-of states in the simple form

$$g_s(\omega) = \omega^2/2\pi^2 \, v_s^3 \qquad (6.38)$$

for frequencies $\omega < \omega_D$ (Debye characteristic frequency) and zero for higher frequencies. Further

$$\frac{\partial N^0(\omega)}{\partial T} = \frac{\hbar \omega}{k_B T^2} \frac{\exp(\hbar\omega/k_B T)}{\{\exp(\hbar\omega/k_B T) - 1\}^2}$$

Substitution for $g_s(\omega)$ and $\partial N^0/\partial T$ in Eq. 6.37 gives

$$\lambda = \frac{1}{3}\frac{1}{2\pi^2} \sum_s \int_0^{\omega_{\max}} \frac{1}{v_s} \hbar \omega^3 \tau(\omega) \frac{(\hbar\omega/k_B T^2) \exp(\hbar\omega/k_B T)}{\{\exp(\hbar\omega/k_B T) - 1\}^2} d\omega \qquad (6.39)$$

Assuming the existence of a single average acoustic branch the expression for λ is further simplified to

$$\lambda = \left(\frac{k_B}{\hbar}\right)^3 \frac{k_B}{2\pi^2 v_s} T^3 \int_0^{\theta_D/T} \frac{x^4 e^x (e^x - 1)^{-2}}{\tau_c^{-1}} dx \qquad (6.40)$$

This is the usual starting point for a large number of thermal conductivity calculations. The existence of a well defined relaxation time having been assumed, the combined relaxation time τ_c can be obtained by the addition of the inverse relaxation times for the different scattering processes

$$\tau_c^{-1} = \sum_i \tau_i^{-1} \tag{6.41}$$

where τ_i refers to the phonon relaxation time for the ith scattering process. Some of the useful phonon relaxation times have been described in Chap. V.

6.5 Normal processes

Although by themselves the momentum conserving (or normal processes) do not contribute to thermal resistance, they do contribute in an indirect way by distributing momentum among all the phonons. This problem was investigated by Callaway (1959) (based upon earlier studies by Klemens, 1951, 1955, 1958 and 1960) who obtained for the thermal conductivity

$$\lambda = \frac{k_B}{2\pi^2 v_s} \left(\frac{k_B T}{\hbar}\right)^3 \left(I_1 + \frac{I_2^2}{I_3}\right) \tag{6.42}$$

The first integral refers to the contribution from the momentum destroying (or umklapp) processes and treats the normal processes at par with them. The second term refers to a correction due to the normal processes. The various integrals are written as

$$I_1 = \int_0^{\theta_D/T} \tau_c \frac{x^4 e^x \, dx}{(e^x - 1)^2}, \qquad I_2 = \int_0^{\theta_D/T} \frac{\tau_c}{\tau_N} \frac{x^4 e^x \, dx}{(e^x - 1)^2}$$

and

$$I_3 = \int_0^{\theta_D/T} \frac{1}{\tau_N}\left(1 - \frac{\tau_c}{\tau_N}\right) \frac{x^4 e^x \, dx}{(e^x - 1)^2} \tag{6.43}$$

In this formulation the normal processes are initially treated at par with other mechanisms which do not conserve crystal momentum. This causes an overestimation of the thermal resistance and is then corrected by the second term in Eq. 6.42. Holland (1963) and Nettleton (1963) have raised objections against this procedure but in spite of its weaknesses it still remains one of the most satisfactory formulations in explaining thermal conductivity data. The correction term is usually small and can therefore be neglected in a wide variety of situations.

The inverse addition of different relaxation times (Eq. 6.41) to obtain the total relaxation time is a good approximation when the various processes are non-interacting and also for elastic scattering with a random distribution of the scatterers. In some cases this may lead to the additivity of the thermal resistivities. However, this is not true in general as the different relaxation times have different frequency dependences.

6.6 Generalization of the Klemens-Callaway expression

Some of the restricting assumptions of the Klemens-Callaway formalism can be eliminated if detailed information is available about the phonon spectrum and the phonon density-of-states distribution. Holland (1963) presented an extension of Callaway's theory in which the total heat carried by phonons is divided into the heat carried by transverse and longitudinal phonons. This model was applied to analyse the thermal conductivity data of silicon and germanium. The strong dispersion of the transverse acoustic branches in these materials requires the development of relaxation times for three-phonon processes which are valid for a certain range of phonon frequencies.

The total lattice thermal conductivity can then be written as (Holland, 1963)

$$\lambda = \lambda_T + \lambda_L \tag{6.44}$$

where

$$\lambda_T = \frac{2}{3} \int_0^{\theta_T/T} \frac{C_T T^3 x^4 \, e^x (e^x - 1)^{-2}}{\tau_T^{-1}} \, dx$$

and

$$\lambda_L = \frac{1}{3} \int_0^{\theta_L/T} \frac{C_L T^3 x^4 \, e^x (e^x - 1)^{-2} \, dx}{\tau_L^{-1}}$$

Here $x = \hbar\omega/k_B T$, $\theta_{T,L} = \hbar\omega_{T,L}/k_B$, and

$$C_{T,L} = \frac{k_B}{2\pi^2 v_{T,L}} \left(\frac{k_B}{\hbar}\right)^3$$

Suffixes T and L refer to transverse and longitudinal branches, respectively. The term λ_T can be divided into contributions from low-frequency transverse modes ($\omega < \omega_1$) and from the transverse modes with frequencies between ω_1 and ω_2, where ω_1 refers to the frequency at which the U-processes start and ω_2 is the highest transverse mode frequency. A more complete expression for τ_U was used in Holland's analysis, which is valid for materials with very disperse transverse acoustic branches This approach was followed by other workers in the analysis of thermal conductivity data (Bhandari and Verma, 1965 a, b).

Kosarev *et al* (1971) presented a generalization of Callaway's thermal conductivity expression which takes account of the differing behaviour of the three polarisation branches. The manner in which the contributions of different polarisation branches were weighted in their derivation was improved upon by Parrott (1971) who derived the following expression for thermal conductivity

$$\lambda = \frac{k_B}{6\pi^2} \left(\frac{k_B T}{\hbar}\right)^3 \left\{ \sum_s \frac{1}{v_s} \int_0^{\theta_s/T} \frac{\tau_N^s \tau_R^s}{\tau_N^s + \tau_R^s} \tilde{N}^0(\tilde{N}^0 + 1) \, x^4 dx \right.$$

$$+ \frac{\left[\sum_s (1/v_s)^3 \int_0^{\theta_{s/T}} \frac{\tau_R^s}{\tau_N^s + \tau_R^s} \tilde{N}^0 (\tilde{N}^0 + 1) x^4 \, dx \right]^2}{\sum_s (1/v_s^5) \int_0^{\theta_{s/T}} \frac{1}{\tau_N^s + \tau_R^s} \tilde{N}^0 (\tilde{N}^0 + 1) x^4 \, dx} \Bigg\} \tag{6.45}$$

s lables the polarisation, $\tilde{N}_0$ is Planck's distribution, τ_R^s the combined relaxation time for the momentum destroying processes, τ_R^N corresponds to the Normal processes, and v_s is the phonon group velocity.

This equation seems to be the appropriate form of the generalized Klemens-Callaway equation and takes into account the separate contributions from the transverse and the longitudinal branches and includes the correction term due to the Normal processes. Bhandari and Rowe (1979) estimated the error that might occur in the theoretical calculation of λ if the appropriate generalization is not taken into consideration.

In some cases it is possible to obtain analytical expressions for λ although, in general, the integrals have to be evaluated numerically. For the case of phonon-phonon scattering combined with point-defect scattering at high temperatures ($T > \theta_D$), it can be shown that (Parrott, 1963 and Abeles, 1963)

$$\frac{\lambda}{\lambda_{pp}} = \frac{1}{1 + \frac{5}{9}k_0} \left\{ \frac{\arctan U}{U} + \frac{(1 - \arctan U/U)^2}{\left(\frac{1+k_0}{5k_0}\right) U^4 - (U^2/3) + 1 - \arctan U/U} \right\} \tag{6.46}$$

where

$$U^2 = \frac{1}{1 + \frac{5}{9}k_0} \left(\frac{9\pi}{2}\right)^{1/3} \frac{\pi^2 \hbar}{k_B^2} \frac{\Gamma\delta}{\theta_D} \lambda_{pp}$$

$$k_0 = \tau_u/\tau_N$$

Parameters Γ and δ have earlier been described in 5.9.2. λ_{pp} is the thermal conductivity when phonon-phonon scattering is the only scattering mechanism present. k_0 is usually adjusted to fit the experimental data. In the limit of weak point-defect scattering ($U \ll 1$) λ is given by (Ambegaokar, 1959)

$$\lambda^{-1} = \lambda_{pp}^{-1} + \left(\frac{9\pi}{2}\right)^{1/3} \frac{\pi^2 \hbar}{3k_B^2} \frac{\Gamma\delta}{\theta_D} \frac{1 + 2k_0 + \frac{25}{21}k_0^2}{\left(1 + \frac{5}{9}k_0\right)^2} \tag{6.47}$$

6.7 Variation of λ with temperature

6.7.1 Introduction

Equation 6.18 summarizes the result obtained by Leibfried and Schlomann (1954) when the thermal conductivity is limited only by three-phonon umklampp processes. The anharmonic forces are related to other physical properties of the solid, such as the Gruneisen constant γ.

This equation determines the variation of λ with temperature if its value at one particular temperature is determined either from theory or experiments. At $T > \theta_D$, λ varies as T^{-1} and rises with decreasing temperature more rapidly than is predicted by the T^{-1} law at somewhat lower temperatures. At temperatures well below the Debye temperature, λ is dominated by the exponential term (see Eq. 6.19), increasing very rapidly and approaching to an infinitely large value at the absolute zero. In real crystals this increase in λ is halted by the finite size of the crystals.

6.7.2 *Effect of crystal boundaries*

Casimir (1938) obtained the following formula for λ, while considering the boundary scattering of phonons

$$\lambda = \frac{1}{3} (\Sigma\, C_j v_j)\, \bar{l}_b \qquad (6.48)$$

C_j and v_j are the contributions to the specific heat and the velocity from the phonons of the jth branch. For a specimen of square cross section $\bar{l}_b = 1.12\, D$, D being the side of the square, while for a cylindrical specimen of radius R, $\bar{l}_b = 2R$.

Berman et al (1953) gave a more detailed theory of boundary scattering and took into consideration different surface conditions. A simple description of the surface is usually employed. A fraction p of the incident particles is assumed to be specularly reflected, as if by a highly polished mirror, and the remainder are scattered diffusely in all directions. In this description $p = 0$ refers to a perfectly rough surface, and $p = 1$ to a perfectly smooth surface. In his calculation, Casimir considered an infinitely long cylinder with perfectly rough walls. Real crystals do not conform to either of these conditions. The crystal surfaces not being perfectly rough some specular reflections of phonons may occur. A surface may appear rough or smooth depending upon the size of the asperities and the wavelengths of the phonons. The effect of specular reflection of phonons and of the finite length of the specimen on boundary scattering have been discussed by Berman *et al.* (1955).

In addition to the scattering of phonons at single crystal boundaries internal boundaries can also cause phonon scattering in polycrystalline or sintered materials.

In his calculations, Casimir neglected all other scattering processes apart from assuming perfect roughness of the surface. In the presence of other scattering processes a common procedure is to take $\tau_b^{-1} = v_s/\bar{l}_b$, where τ_b is the relaxation time due to boundary scattering, v_s is the average sound velocity and $\bar{l}_b$ corresponds to a characteristic crystal dimension. However, this procedure is not quite accurate as the phonon distribution function N_q may have different values at the surface and in regions far from the surface, and in any rigorous calculation this fact should be taken into consideration.

6.7.3 *Boundary scattering at high temperature*

Boundary scattering is usually thought of as a low-temperature phenomenon. However, in highly disordered materials it may become appreciable even at high temperatures (Goldsmid and Penn, 1968). Point-defect scattering is very effective (ω^4 dependence) for short-wavelength phonons. In highly disordered materials the short-wavelength phonons are scattered effectively and most of heat is carried by long-wavelength phonons. These phonons can then be scattered by the grain-boundaries. Herring (1954) had earlier discussed the size dependence of the thermal conductivity and Geballe and Hull (1955) obtained an appreciable size dependence in relatively thick specimens of Ge well beyond the conductivity maximum. Experiments by Savvides and Goldsmid (1973) on the thermal conductivity of thin specimens of silicon at 197 and 293 K showed a pronounced decrease in λ due to boundary scattering which was further enhanced in samples irradiated with neutrons (Savvides and Goldsmid, 1974).

Parrott (1969) further explored the consequences of grain-boundary scattering in fine-grained material. Taking the high temperature limit and replacing $(e^x - 1)$ by $x(x = \hbar\omega/k_B T)$, it is possible to obtain simplified formulae. The relaxation times for the umklapp and normal processes are taken to be $\tau_u^{-1} = A_U x^2$ and $\tau_N^{-1} = k_0 A_u x^2$, respectively. Taking λ_0 as the thermal conductivity of a virtual perfect crystal, thermal conductivity can be written as (Parrott, 1969, Meddins and Parrott, 1976, and Bhandari and Rowe, 1978)

$$\lambda/\lambda_0 = \left(1 + \frac{5k_0}{9}\right)^{-1} \left[L_2(A, C) + \frac{\left(\frac{k_0}{1+k_0}\right) L_4^2(A, C)}{\left\{\frac{1}{5} - \left(\frac{k_0}{1+k_0}\right) L_6(A, C)\right\}} \right] \quad (6.49)$$

where

$$L_n(A, C) = \int_0^1 \frac{x^n \, dx}{Ax^4 + x^2 + C}$$

The parameters A nnd C are defined as

$$A = \frac{\pi \Gamma \Omega_0 \lambda_0 \omega_D}{2v_S^2 k_B(1 + 5k_0/9)}, \qquad C = \frac{2\pi^2 \lambda_0 v_S}{k_B L \omega_D^3 (1 + 5k_0/9)} \quad (6.50)$$

L refers to the crystal dimension. The parameter A describes the effect of disorder and C the effect of grain boundaries. Evidently

$$\lambda (A = 0, C = 0) = \lambda_0.$$

These formulae, although oversimplified due to the approximate manner in which the N-processes are included, have proved to be very useful in describing the thermal conductivity behaviour of fine-grained semiconductor alloys. Applications of these equations to "real" materials are described in

114 THERMAL CONDUCTION IN SEMICONDUCTORS

Chap. VII where the effect of doping has also been included in the calculation.

The general pattern of the variation of λ with temperature can then be described by taking into consideration boundary scattering of phonons along with three-phonon scattering and is shown in Fig. 6.2. A T^3 varia-

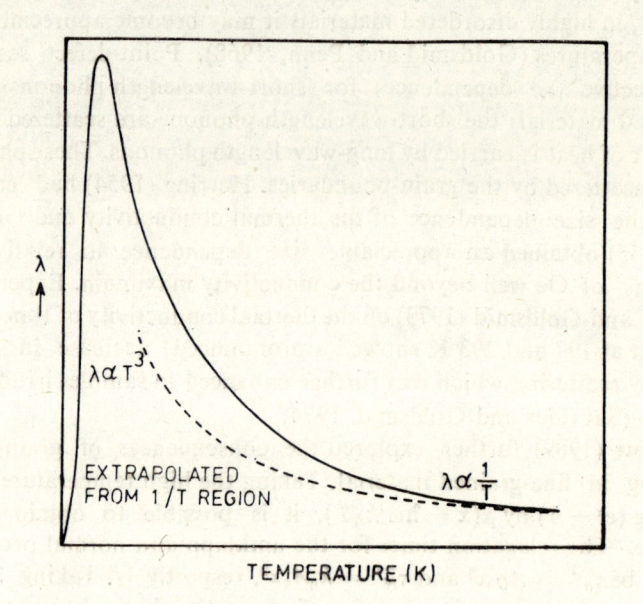

Fig. 6.2 Schematic plot of general theoretical behaviour of thermal conductivity for a perfect large crystal (after Drabble and Goldsmid, 1961).

tion at low temperatures and a T^{-1} variation at temperatures well above the Debye temperature is predicted on the basis of these considerations.

6.8 The 1/T law

Although the $1/T$ variation of thermal conductivity at high temperatures gives a reasonably good description of the observed data there are deviations from this law. For a number of III-V semiconductors λ shows a more rapid variation than is given by the T^{-1} law (Steigmeier, 1969). Slack (1972) pointed out that a change in the volume with temperature can produce significant departures from the T^{-1} law. This is apparent from Eq. 6.18 where the parameters a, θ_D and γ are volume dependent. If measurements are made at constant volume, the results are expected to be closer to the T^{-1} law. The temperature dependence of λ_p (thermal conductivity at constant pressure) and λ_v (thermal conductivity at constant volume) are described as

$$\lambda_P = A'/T^\epsilon$$
$$\lambda_v = B'/T^m$$

(6.51)

Obviously, $m=1$ for three-phonon processes. A' and B' depend upon pressure and volume, respectively but are independent of temperature.

Ranninger (1965) pointed out that if the temperature variation of volume is taken into consideration, then $\epsilon > m$ and the measured ϵ values in rare gas crystals can be explained on the basis of these considerations. However, the agreement is not always good for other materials including a large number of semiconductors. Pomeranchuk (1941) suggested that four-phonon processes could be important in many cases and could possibly explain deviations of ϵ from unity.

It was pointed out (Ecsedy and Klemens, 1975, and Klemens and Ecsedy, 1976) that four-phonon processes might be too weak to explain significant departures from the T^{-1} law and that the three-phonon processes including both the acoustic and the optic phonons may explain such departures (see Appendix B).

6.9 Doped semiconductors

The thermal conductivity of doped semiconductors requires separate evaluation of the electronic and the lattice contributions. Apart from acting as carriers of heat, electrons and holes act as scattering centres for phonons and thus cause a reduction in lattice thermal conductivity. The phonon-electron scattering and the corresponding relaxation times have been described in Chap. V. In metals the effect of this scattering is significant in limiting the phonon mean-free-path. However, in semiconductors the electron concentration being several orders of magnitude smaller, the reduction in λ is relatively less than that in metals. Steigmeier and Abeles (1964) pointed out that phonon-electron scattering may be significant in materials which show strong disorder scattering. At temperatures well above the Debye temperature the integrand appearing in the thermal conductivity equation (such as Eq. 6.40) can be shown to be proportional to $\tau_c z^2$ where $z = \omega/\omega_D$, and τ_c is the combined relaxation time obtained as $\tau_c^{-1} = \sum_i \tau_i^{-1}$. A plot of $\tau_c z^2$ against z for the different combinations of scattering mechanisms illustrates the effectiveness of the electron-phonon scarttering with and without point-defect scattering (Fig. 6.3). The effect of phonon-electron scattering appears to be enhanced in the presence of point-defect scattering. This is somewhat analogous to the increase in the effectiveness of the boundary scattering in highly disordered material. Both the boundary scattering and the phonon-electron scattering limit more effectively the mean-free-paths of long-wavelength phonons and are bound to be more effective when short-wavelength phonons are already scattered by other processes such as disorder scattering. The difference in the behaviour of undoped and doped silicon-germanium alloys with regard to grain-boundary scattering may be understood on these lines (Chap. VII). The effect of doping on the

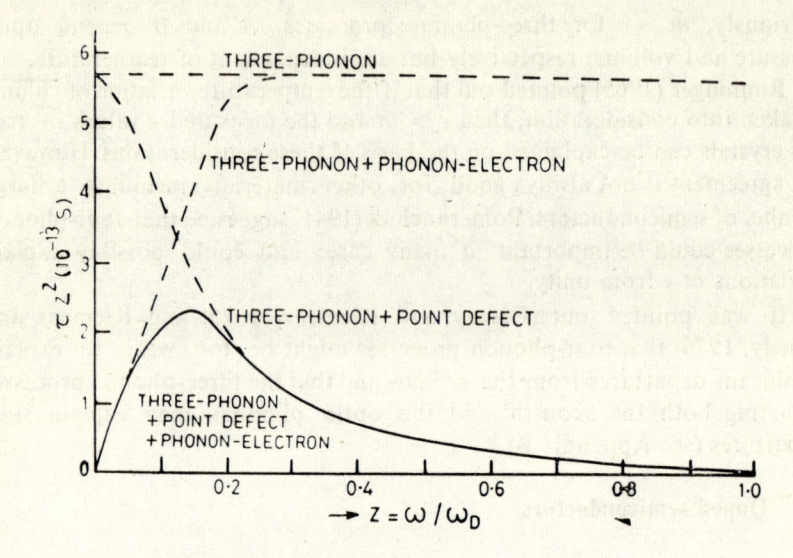

Fig. 6.3 Integrand in the thermal conductivity expression against reduced phonon frequency (ω/ω_D) for different scattering processes (after Steigmeier and Abeles, 1964).

high-and-low temperature thermal conductivity of various semiconductors will be described in the next two chapters.

6.10 Imperfect crystals

The high-temperature T^{-1} dependence and the low-temperature T^3 dependence of the lattice thermal conductivity can be explained on the basis of umklapp three-phonon scattering and boundary scattering. The comparison between theoretical and experimental results presented in Sec. 6.3 shows good agreement at $T = \theta_D$ for a large number of perfect non-metallic crystals. However, imperfections are invariably present and even in pure crystals the presence of various isotopes can drastically influence the thermal conductivity. The high- and low-temperature behaviour may still be described by the T^{-1} and T^3 laws but for the intermediate range of temperatures, particularly in the region of the conductivity maximum, the additional scattering by impurities has to be taken into consideration. Figure 6.4 shows $\lambda - T$ curves for two samples of Ge; one normal and the other in which the proportion of the isotope Ge^{74} has been increased to 96%. At high temperatures, the difference between the two curves becomes independent of temperature. Ambegaokar (1959) pointed out that these observations are in good agreement with the theory of phonon scattering by point imperfections.

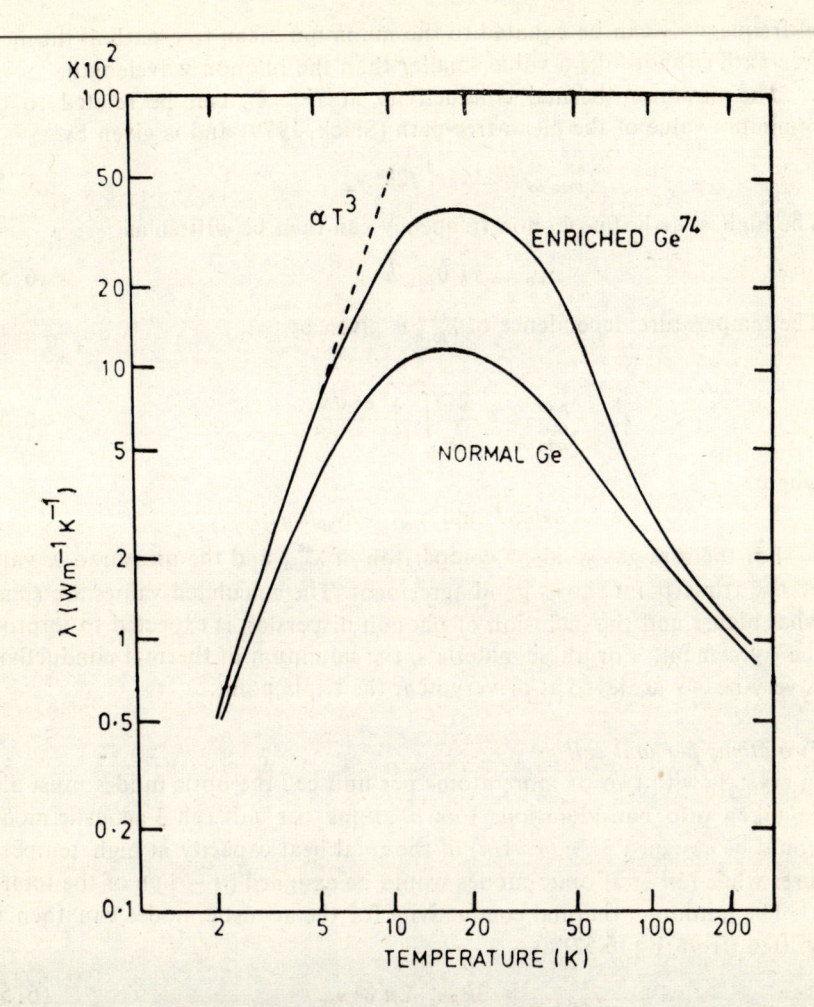

Fig. 6.4 Thermal conductivity of normal Ge and enriched isotope Ge^{74} (after Geballe and Hull, 1958).

6.11 Minimum of thermal conductivity

Slack (1979) has recently considered the concept of a minimum of thermal conductivity for non-metallic solids which was earlier discussed by Roufosse and Klemens (1974). The basic idea behind this study is to determine the minimum possible thermal conductivity of a material and its temperature dependence. This subject is of interest in the calculation of maximum efficiency in thermoelectric devices (see Chap. XI).

The simplest situation is that of the crystals with one atom per unit cell (such as the rare-gas crystals) where only acoustic phonons are present. In a simple calculation, a Debye model is used and all the phonons are assumed to have the same group velocity v_s. The wavelength v_s/ν for a phonon of frequency ν can be equated to the minimum mean-free-path as the mean free-path cannot take a value smaller than the phonon wavelength.

The minimum thermal conductivity at $T = \theta_D$ can be related to the minimum value of the mean-free-path (Slack, 1979) and is given by

$$\lambda_{\min \infty}^{ac} = 3k_B v_s^2 / 2\delta^3 \, \nu_{ac} \tag{6.52}$$

The highest acoustic-phonon frequency can then be written as

$$\nu_{ac} = k_B \tilde{\theta}_{D\infty}/h \tag{6.53}$$

The temperature dependence of $\lambda_{\min}^{ac}$ is given by

$$\frac{\lambda_{\min}^{ac}}{\lambda_{\min \infty}^{ac}} = \frac{2}{x_{ac}^2} \int_0^{x_{ac}} \frac{x^3 e^x \, dx}{(e^x - 1)^2} \tag{6.54}$$

where

$$x = h\nu/k_B T \text{ and } x_{ac} = h\nu_{ac}/k_B T$$

For the rare-gas solids, a comparison of $\lambda_{\min}^{ac}$ and the measured λ value at the triple-point shows good agreement. The calculated values are somewhat higher and the inclusion of phonon dispersion is expected to improve the agreement. For these materials, the minimum of thermal conductivity is very nearly achieved at or very near the triple point.

Two atoms per unit cell
In crystals with two or more atoms per unit cell the optic modes must also be taken into consideration. For n atoms per unit cell 3 acoustic modes would be assigned $3/3n$ ($= 1/n$) of the total heat capacity at high temperatures while $(3n - 3)$ optic modes would be assigned $(n - 1)/n$ of the total.

The minimum thermal conductivity for the acoustic modes can then be written (from Eq. 6.52) as

$$\lambda_{\min \infty}^{ac} = 3k_B v_s^2 / 2n \, \delta^3 \, \nu_{ac} \tag{6.55}$$

For an optic mode of frequency ν_0 the minimum mean-free-path is δ. The velocity of propagation will effectively be given by $v = \delta\nu_0$, and for a single set of three optic branches

$$\lambda_{\min \infty}^{op} = \frac{k_B \nu_0}{n\delta} \tag{6.56}$$

For $n \geq 3$, a sum over all optic branches is taken

$$\lambda_{\min \infty}^{op} = \frac{k_B}{n\delta} \sum_{i=1}^{n-1} \nu_{0i} \tag{6.57}$$

The temperature dependence of λ can be obtained in an approximate way by assuming that the specific heat can be represented by an Einstein oscillator of frequency ν_0.

The total minimum thermal conductivity is obtained by taking a sum of acoustic and optic phonon contributions

LATTICE THERMAL CONDUCTIVITY

Table 6.1
The minimum thermal conductivity at the melting point for crystals with $n = 2$ (after Slack, 1979)

Crystal	δ (10^{-10} m)	ν_A (THz)	ν_0 (THz)	v (10^3 ms^{-1})	$\lambda_{min\,\infty}^{ac}$	$\lambda_{min\,\infty}^{op}$	λ_{min}^{total} (W/mK)	λ_{meas}
Si	2.73	8.23	15.3	6.36	2.501	0.387	2.888	13.9
Ge	2.85	4.90	9.00	3.75	1.287	0.218	1.505	10.0
InSb	3.25	2.81	5.71	2.53	0.688	0.121	0.809	5.7
CdTe	3.26	2.50	5.08	2.26	0.612	0.108	0.720	1.1
LiF	2.09	10.4	19.7	4.21	1.935	0.651	2.586	2.7
NaBr	2.77	3.13	6.22	1.89	0.559	0.155	0.714	0.54
KCl	3.26	3.58	4.10	2.19	0.402	0.087	0.489	1.2
KBr	3.42	2.44	5.00	1.67	0.297	0.101	0.398	0.47
KI	3.66	1.81	4.3	1.37	0.219	0.081	0.310	0.35
AgBr	2.96	2.29	4.10	1.11	0.216	0.096	0.312	0.41
PbTe	3.29	2.19	3.4	1.39	0.257	0.071	0.328	0.48

120   THERMAL CONDUCTION IN SEMICONDUCTORS

$$\lambda_{\min}^{\text{total}} = \lambda_{\min}^{\text{ac}} + \lambda_{\min}^{\text{op}} \tag{6.58}$$

Table 6.1 presents the calculated $\lambda_{\min}^{\text{total}}$ for a large number of non-metallic crystals with $n = 2$. The observed values refer to the values obtained by extrapolating high-temperature values to the melting point. The δ values were calculated from the molar volumes of the solids at their melting points. The observed values refer to the lattice contribution only.

It is clear from the table that, for the rocksalt structure, the measured λ at the melting point is about 1 to 2 times the calculated minimum. For adamantine crystals this ratio is 1.5 to 7 times the calculated minimum. For these crystals phonon-phonon scattering has not reached its maximum effectiveness at the melting point and there is, in principle, scope for further reduction.

The next step is to determine how close the calculated minimum approaches the measured values of λ. Silicon-germanium alloys have a large mass-difference scattering due to the widely differing atomic masses of the constituents and therefore the conductivity of Si-Ge alloys is considerably lower than that of Si or Ge. Doping causes an additional reduction in λ and in heavily doped semiconductors it is primarily reduced due to additional scattering of the phonons by electrons. Even then the measured λ is twice the calculated minimum at 300 K. Similar results have been obtained for the mixed crystals of PbTe, InSb and CdTe. This suggests that there is further scope of reducing λ in these materials.

The thermal conductivity of amorphous solids may be of particular interest in this context as these materials are likely to have thermal-conductivity values approaching the calculated minimum. However, the thermal conductivity of amorphous germanium, as measured by Nath and Chopra (1974), is an order of magnitude greater than the predicted minimum.

Goldsmid et al. (1983) developed a technique of measuring the thermal conductivity of thin films and made measurements on films of amorphous silicon and amorphous germanium (Goldsmid and Paul, 1983). The measured λ values for both the materials were 2.6 Wm^{-1} K^{-1} with an estimated error of $\pm$ 15% while the theoretical minimum values of thermal conductivity were 2.9 Wm^{-1} K^{-1} and 1.5 Wm^{-1} K^{-1}, respectively (Slack 1979). The technique of measurement used by Goldsmid et al. being more suitable for poorly conducting materials, their results are closer to the theoretical values for the two cases mentioned.

6.12 Method of Guyer and Krumhansl

A different approach towards the calculation of thermal conductivity was given by Guyer and Krumhansl (1966). This approach also takes account of the special features of the N-processes. The thermal conductivity is written as

$$\lambda = \frac{1}{3} C_v v^2 \frac{\langle \tau_R \rangle}{\langle \tau_R \rangle + \langle \tau_N \rangle} \left[\langle \tau_N \rangle + \frac{1}{\langle \tau_R^{-1} \rangle} \right] \quad (6.59)$$

τ_N and τ_R refer to the phonon relaxation times for the normal and resistive (momentum destroying) processes. For strong N-processes $\tau_N^{-1} \gg \tau_R^{-1}$, and the expression leads to the second term of the Callaway equation. On the other hand for very weak N-processes $\tau_N^{-1} \ll \tau_R^{-1}$, and the usual expression for thermal conductivity is obtained. For the intermediate range of the relative values of τ_N and τ_R the analysis gives results similar to those obtained from the Callaway expression but there are differences when $\langle \tau_R \rangle \simeq \langle \tau_R^{-1} \rangle^{-1}$ (Day, 1970).

The case of boundary scattering is particularly well handled by this method. When N-processes are strong, phenomena of essentially hydrodynamic nature (like Poiseuille flow) are predicted. A flow of phonons not affected by the boundary scattering, except in a layer within about $\tau_N v$ of the surface, is possible in this situation. This leads to a reduction in the boundary scattering rate compared with the case with no N-processes. In this situation, additional scattering can lead to an increase in the mean-free-path. This type of situation has so far been observed only in solid helium (Hogan et al., 1969).

6.13 Other carriers of heat

6.13.1 Introduction
In most situations, the measured thermal conductivity of semiconductors can be analysed in terms of contributions from phonons, electrons (or holes) and, in the intrinsic conduction range, one from electron-hole pairs. However, in certain cases other mechanisms of heat transfer may become significant and must therefore be taken into consideration. Heat conduction by photons and magnons have been investigated in some detail and there have been suggestions about a possible excitonic contribution to λ. Magnon thermal conductivity shall be discussed in the section on magnetic semiconductors in Chap. X.

6.13.2 Photon thermal conductivity
Thermal radiations corresponding to a mean temperature are present within every material and in thermal equilibrium these radiations are blackbody radiations. If the material is transparent for a particular photon it can pass through the material undisturbed. If not, photons will diffuse through the material in a manner similar to the flow of phonons down a temperature gradient. Genzel (1953) was the first to obtain a theoretical expression for the photon thermal conductivity. Devyatkova et al. (1959) extended the theory to uniaxial crystals. The subject has been reviewed by Men' and Sergeev (1973).

Photon thermal conductivity can be written as (Genzel, 1953)

$$\lambda_{\text{photon}} = \frac{16}{3}\sigma_0 n^2 T^3 \alpha^{-1} \qquad (6.60)$$

where σ_0 is the Stefan-Boltzmann constant, n is the refractive-index and α is the absorption coefficient.

This expression is quite similar to the general thermal conductivity equation, $\lambda = (1/3)\, C_v v^2 \tau$, where C_v is replaced by the specific heat of the photon gas, $16\, N^3 T^3 \sigma_0/c$. Further, $v = c/N$, where c is the velocity of light and $\tau v = 1/\alpha$.

Photon thermal conductivity has been observed in a number of materials and is most likely to be significant in materials with large energy gaps and low lattice thermal conductivity; tellurium, selenium, silicon–germanium alloys are some such materials. The lead chalcogenides also show an appreciable photon thermal conductivity. Photon thermal conduction in silicon-germanium alloys is further discussed in Chap. VII.

Excitonic thermal conductivity

Heat transport by excitons was earlier suggested by Ioffe (1956) and this contribution was thought to be appreciable in lead telluride (Devyatkova, 1957). Pavlov and Eshpulatov (1977) derived an expression for the excitonic thermal conductivity taking into account the scattering of excitons by phonons, and also studied the effect of exciton-drag on thermal conductivity. However, so far no conclusive evidence has been obtained to confirm its existence.

Appendix A

Improved Variational Principles for Lattice Thermal Conductivity

The variational method described earlier in Chap. III and in this chapter (Sec. 6.2) gives only a lower estimate of the transport coefficient. Jensen *et al.* (1969) described the possibility of obtaining a variety of bounds. Benin (1970) pointed out that the approach of Jensen and co workers was valid only when the linearized collision operator for the system possessed a bounded eigenvalue spectrum, and that this method was applicable in the case of anharmonically interacting phonons. For collision operators with a bounded spectrum Benin obtained a sequence of functions $\lambda_m^{\leq}$, $m = 0, 1, 2, \ldots$, which give lower bounds of λ. The first term $\lambda_0^{\leq}$ yields the lower bound of Ziman's. The infinite sequence is found to converge monotonically towards the exact transport coefficient.

Following Arthurs' (1970) suggestion on the complementary variational principles it is possible to obtain both upper and lower bounds on the same transport coefficients. Srivastava (1975) showed that a sequence of upper bounds for the coefficient could be obtained. The sequence is shown to converge and approach monotonically the exact λ from above. The two sequences $\{\lambda_m^\leq\}$ of lower bound and $\{\lambda_n^\geq\}$ of upper bounds greatly increase the usefulness of the method of complementary variational principles (CVR's) and λ can now be confined to a range $(\lambda_n^\geq - \lambda_m^\leq)$ around λ.

There are various methods of deriving CVPs and all these methods are connected, and in some cases, equivalent. The sequences of maximum and minimum variational principles for the lattice thermal conductivity, using the Schwarz inequality and those obtained using canonical Euler equations, are shown to be equivalent (Srivastava, 1976). Srivastava and Hamilton (1978) reviewed various methods of arriving at CVPs and discussed their application to the theory of thermal conductivity.

Appendix B

Effect of Thermal Expansion and Scattering by Optic Phonons

An account of deviations from the $1/T$ law and some attempts to explain it were earlier given in Sec. 6.8. Most of the thermal conductivity measurements are made at constant pressure and as the volume changes with temperature it is natural to expect a somewhat different dependence of λ measured at constant volume. The slopes of $\ln \lambda$ versus $\ln T$ curves for $T \geqslant \tilde{\theta}_{D\infty}$ at constant pressure and constant volume may take values ϵ and m given by (Slack, 1979)

$$\epsilon = -\left(\frac{\partial \ln \lambda}{\partial \ln T}\right)_P \tag{B.1}$$

$$m = -\left(\frac{\partial \ln \lambda}{\partial \ln T}\right)_V$$

For $T \geqslant \tilde{\theta}_{D\infty}$, the temperature dependences of λ_P and λ_V are described by Eq. 6.51. For acoustic three-phonon processes the theoretical value of m is equal to unity. It can be shown that

$$\epsilon = m + gT\left(\frac{\partial \ln V}{\partial T}\right)_P \tag{B.2}$$

where g is given by

$$g = -\left(\frac{\partial \ln \lambda}{\partial \ln V}\right)_T \tag{B.3}$$

Equating 3α to $(\partial \ln V/\partial T)_P$, where α is the coefficient of linear thermal expansion, one obtains (Slack, 1979)

$$\epsilon = m + \eta_{\text{th}} \tag{B.4}$$

where

$$\eta_{\text{th}} = 3\alpha gT$$

For the rare-gas crystals there is a very good agreement between the theoretical and observed values of ϵ.

Application of this simple formulation to a number of compounds of rocksalt structure shows that for compounds with a mass-ratio (of the components) close to unity, the agreement between theoretical and observed values of ϵ is good but not so for compounds with a ratio which is far from unity.

The effect of thermal expansion on the thermal conductivity has also been discussed by Srivastava (1981) through its effect on the Gruneisen constant which becomes temperature-dependent (Yates, 1972, and Soma, 1977).

Ecsedy and Klemens (1975 and 1976) suggested that three-phonon processes involving optic phonons may explain m values greater than 1. In the high-temperature range where only acoustic phonon scattering is significant the thermal resistivity is written as

$$W = 1/\lambda = aT \tag{B.5}$$

A model in which only acoustic phonons act as heat carriers with both acoustic and optic phonons acting as scattering agents has been described (Slack, 1979) to explain observed ϵ values. In a simple analysis we may take $n = 2$ and assume an optic phonon branch with a single frequency ν_{op}. For acoustic phonons a Debye spectrum is assumed with a maximum frequency equal to $k_B \tilde{\theta}_{D\infty}/h$. With $\theta_{D,op} = h\nu_{op}/k_B$ and $\theta_{D,op} > \tilde{\theta}_{D\infty}$ and further assuming that the scattering of acoustic phonons by other acoustic phonons or by optic phonons is proportional to the average thermal energy present in each branch the thermal resistance is shown to be

$$W \simeq \frac{aT}{(1+S_{op})}\left[1 + S_{op}\left(\frac{x}{e^x - 1}\right)\right] \tag{B.6}$$

where $x = h\nu_{op}/k_B T$ and S_{op} = relative strength of the optic-acoustic phonon interaction.

In the high-temperature limit, $x \to 0$ and Eq. B.6 reduces to Eq. B.5 irrespective of the value of S_{op}. The logarithmic derivative of Eq. B.6 gives

$$m = \left(\frac{\partial \ln W}{\partial \ln T}\right) = 1 + \eta_{op} \tag{B.7}$$

η_{op} is the contribution to ϵ from optic phonons and is given by

$$\eta_{op} = S_{op}z(z + x - 1)/(S_{op}z + 1) \tag{B.8}$$

where

$$x = \theta_{D,op}/T \quad \text{and} \quad z = x/(e^x - 1)$$

Adding various contributions to ϵ we obtain

$$\epsilon = 1 + \eta_{th} + \eta_{op} \tag{B.9}$$

The results of calculations along with the measured values of ϵ are shown in Fig. 6.5 and Fig. 6.6 for Si and CdTe. In both cases S_{op} is given a value equal to 1.2. From the results it is clear that for $T > \theta_{D\infty}$, the measured λ

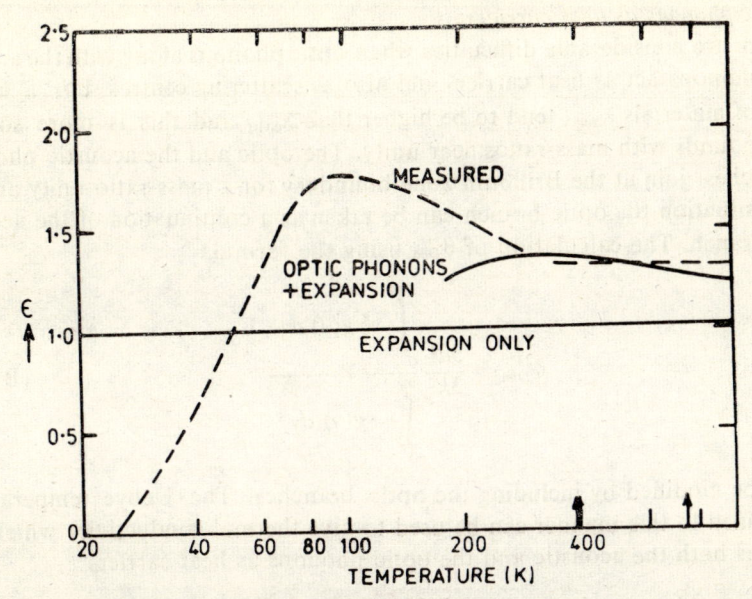

Fig. 6.5 Temperature variation of the exponential factor, ϵ, for silicon. Solid line shows the calculated values (after Slack, 1979).

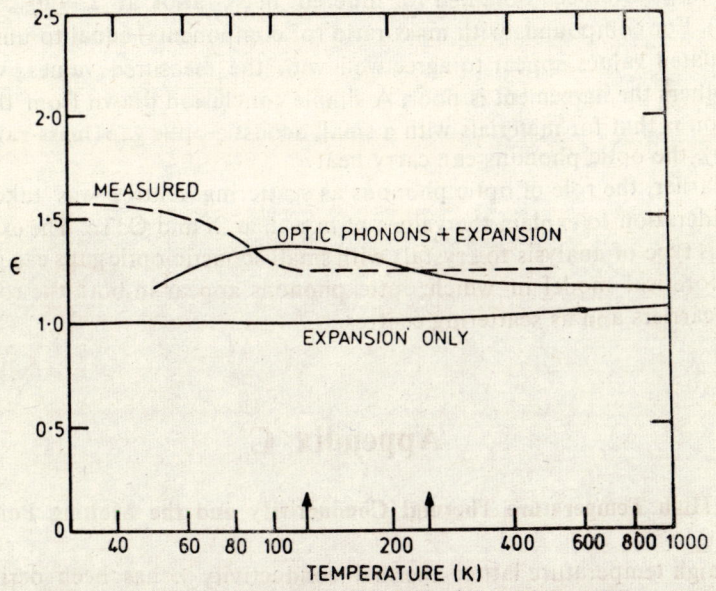

Fig. 6.6 ϵ versus temperature for CdTe (after Slack, 1979).

can be reasonably explained on the basis of a simple model which takes into consideration three-phonon processes, thermal expansion and scattering by optic phonons. For more complex systems ($n > 2$), a more complex model may be needed but the basic features are expected to remain the same.

Optic phonons as heat carriers

There are considerable difficulties when optic phonons along with the acoustic phonons act as heat carriers and also as scattering centres. For a number of materials λ_{meas} tend to be higher than λ_{calc} and this is more so for compounds with mass-ratios near unity. The optic and the acoustic phonon branches join at the Brillouin zone boundary for a mass-ratio unity and in this situation the optic branch can be taken as a continuation of the acoustic branch. The calculation of $\theta_{D\infty}$ using the formula

$$\theta_{D\infty}^2 = \frac{5h^2}{3k_B^2} \frac{\int_0^\infty v^2 g(v)\, dv}{\int_0^\infty g(v)\, dv} \tag{B.10}$$

can be modified by including the optic branches. The Debye temperature obtained in this manner can be used to give thermal conductivity which includes both the acoustic and the optic phonons as heat carriers

$$\lambda = Bn^{1/3}\overline{M}\,\delta\theta_{D\infty}^2\,\gamma_\infty^{-2} \tag{B.11}$$

γ_∞ defers from $\tilde{\gamma}_\infty$ in the same manner as $\theta_{D\infty}$ defers from $\tilde{\theta}_{D\infty}$. Values of $\lambda_{\text{meas}}/\lambda_{\text{calc}}$ have been obtained for different mass-ratios at $T = \tilde{\theta}_{D\infty}$ (Slack, 1979). For compounds with mass-ratio (of components) equal to unity, the calculated values appear to agree well with the measured values, whereas for others the agreement is poor. A simple conclusion drawn from this discussion is that for materials with a small acoustic-optic gap (mass-ratio near unity), the optic phonons can carry heat.

Earlier, the role of optic phonons as scattering centres was taken into consideration to explain the values of $m > 1$ in Si and CdTe. The extension of this type of analysis to crystals with small acoustic-optic gaps can provide a theoretical model in which optic phonons appear in both the roles—as heat carriers and as scattering centres.

Appendix C

High Temperature Thermal Conductivity and the Melting Point

The high temperature lattice thermal conductivity λ has been derived in several forms by various authors. A formula suggested by Dugdale and Macdonald (1955) relates λ to compressibility χ, sound velocity, v and r_0 (which is the nearest-neighbour distance) and the Gruneisen parameter γ

$$\lambda = r_0 v/3\chi\gamma^2 T \tag{C.1}$$

v can be related to and density ρ as $v = (\rho\chi)^{-1/2}$.

An alternative expression for λ can be obtained from these equations (Lawson, 1957)

$$\lambda = r_0/3\rho^{1/2}\chi^{3/2}\gamma^2 T \qquad (C.2)$$

Using the Lindemann melting rule, Keyes (1959) eliminated χ in favour of the melting temperature T_m of the material. According to this rule melting takes place when the amplitude of the thermal vibrations of the atoms reaches some fraction ϵ' of the interatomic distance; ϵ' is taken to be the same for all solids. Keyes obtained the following relationship

$$\lambda T = \frac{B_1 T_m^{3/2} \rho^{2/3}}{A^{7/6}} \qquad (C.3)$$

where
$$B_1 = \frac{R^{3/2}}{3\gamma^2 \epsilon'^3 N_L^{1/3}}$$

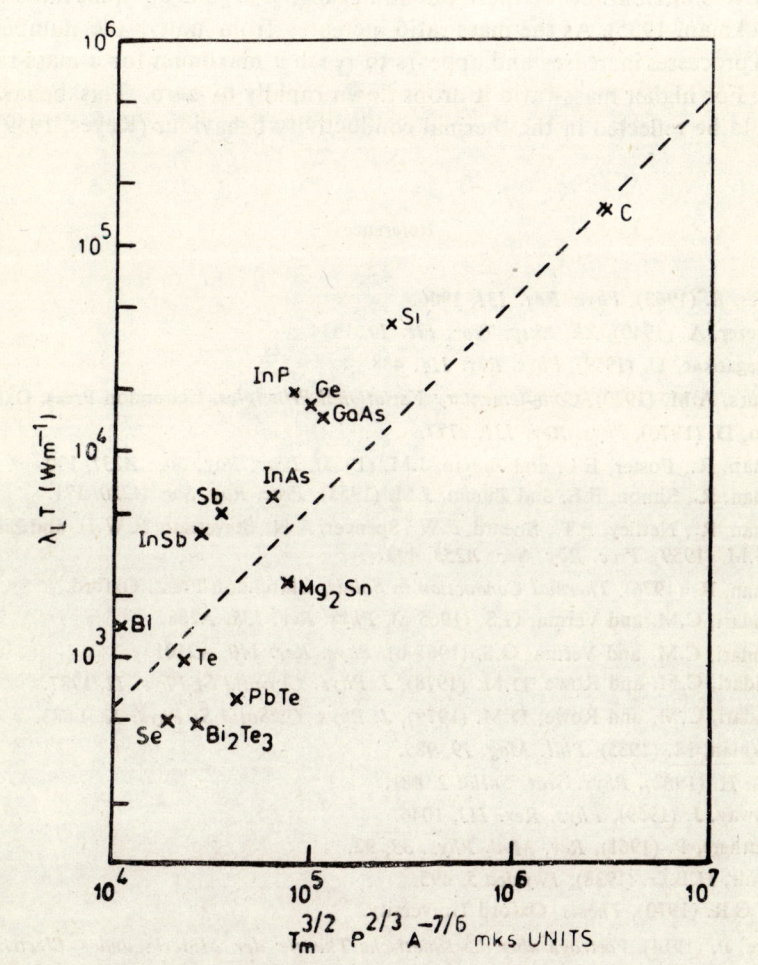

Fig. 6.7 Relationship between $\lambda_L T$ and the melting point for various semiconductors (after Drabble and Goldsmid, 1961)

Here R is the gas constant, A the atomic mass and N_L the Loschmidt number. According to Keyes, B_1 is almost the same for a variety of solids. A plot of λT versus $T_m^{3/2} \rho^{2/3} A^{-7/6}$ is given in Fig. 6.7. The formula is satisfied within an order of magnitude for materials with their λT values spread over about four orders of magnitude. In general, this formula is more reliable for comparing the thermal conductivities of materials belonging to the same class. The value of B_1 is very similar for a particular class of materials while it may differ significantly between two separate classes.

The departure of the observed λT from the predicted value could be related to a variation of mass-ratio among different types of atoms in a solid. Eucken and Kuhn (1928) studied a number of alkali halide crystals to investigate the effect of a change in the mass-ratio. The umklapp-processes responsible for limiting λ involve acoustic as well as optic phonons, and the relative contributions of these phonon branches depend upon the mass-ratio (Blackman, 1935). As the mass-ratio increases from unity, the number of such processes increases and appears to reach a maximum for a mass-ratio of 3. For higher mass-ratio it drops down rapidly to zero. This behaviour should be reflected in the thermal conductivity behaviour (Keyes, 1959).

References

Abeles, B. (1963), *Phys. Rev. 131*, 1906.
Akhiezer, A. (1940), *Zh. Eksp. Teor. Fiz. 10*, 1354.
Ambegaokar, U. (1959), *Phys. Rev. 114*. 488.
Arthurs, A.M. (1970), *Complementary Variational Principles*, Clarendon Press, Oxford.
Benin, D. (1970), *Phys. Rev. 131*, 2777.
Berman, R., Foster, E.L. and Ziman, J.M. (1955), *Proc. Roy. Soc. A231*, 130.
Berman, R., Simon, F.E. and Ziman, J.M. (1953), *Proc. Roy. Soc. A220*. 171.
Berman, R., Nettley, P.T., Sheard, F.W., Spencer, A.N., Stevenson, R.W.H. and Ziman, J.M. (1959), *Proc. Roy. Soc. A253*, 403.
Berman, R. (1976), *Thermal Conduction in Solids*, Clarendon Press, Oxford.
Bhandari, C.M. and Verma, G.S. (1965 a), *Phys. Rev. 138*, A288.
Bhandari, C.M. and Verma, G.S. (1965 b), *Phys. Rev. 140*, A2101.
Bhandari, C.M. and Rowe, D.M. (1978), *J. Phys. C: Solid St Phys. 11*, 1787.
Bhandari, C.M. and Rowe, D.M. (1979), *J. Phys. C: Solid St Phys. 12*, L883.
Blackman, M. (1935), *Phil. Mag. 19*, 989.
Bross, H. (1962), *Phys. Stat. Solidi 2*, 481.
Callaway, J. (1959), *Phys. Rev. 113*, 1046.
Carruthers, P. (1961), *Rev. Mod. Phys. 33*, 92.
Casimir, H.B.G. (1938), *Physica 5*, 495.
Day, C.R. (1970), *Thesis*, Oxford University.
Debye, P. (1914), *Vortrage uber die Kinetische Theorie der Materie under Electrizitat*, B.G. Teubner, Berlin, p43.
Devyatkova, E.D. (1957), *Sov. Phys.—Tech. Phys. 2*, 414.

Devyatkova, E.D., Moizhes, V. Ya. and Smirnov, I.A. (1959), *Fiz. Tveraoga Tela 1*, 613.
Drabble, J.R. and Goldsmid, H.J. (1961), *Thermal Conduction in Semiconductors*, Pergamon Press, London, Ch. 5.
Dugdale, J.S. and MacDonald, D.K.C. (1955), *Phys. Rev. 98*, 1751.
Erdos, P. (1965), *Phys. Rev. 138*, A1200.
Ecsedy, D.J. and Klemens, P.G. (1975), *Bull. Amer. Phys. Soc. 20*, 356.
Eucken, A. and Kuhn, E. (1928), *Z. Physik Chem. Frankfurt 134*, 193.
Geballe, T.H. and Hull, G.W. (1955), *Conf. de Physique des Basses Temp.*, Paris, p. 460, Annexe 1955-3, *Supplement au Bulletin de l' Institut Int. du Froid*, Paris 17.
Geballe, T.H. and Hull, G.W. (1958), *Phys. Rev. 110*, 773.
Genzel, L. (1953), *Z. Physik 135*, 177.
Goldsmid, H.J., Kaila, M.M. and Paul, G.L (1983), *Phys. Stat. Solidi 76*, K 31.
Goldsmid, H.J. and Paul, G.L. (1983), *Thin Sol. films 103*, L 47.
Goldsmid, H.J. and Penn, A.W. (1968), *Phys. Lett. 27A*, 523.
Guthrie, G.L. (1971), *Phys. Rev. B3*, 3573.
Guyer, R.A. and Krumhansl, J.A. (1966), *Phys. Rev. 148*, 766.
Hamilton, R.A.H. and Parrott, J.E. (1969), *Phys. Rev. 178*, 1284; also *Phys. Lett. A29*, 556 (1969).
Herring, C. (1954), *Phys. Rev. 95*, 954.
Hogan, E.M., Guyer, R.A. and Fairbank, H.A. (1969), *Phys. Rev. 185*, 356.
Holland, M.G. (1963), *Phys. Rev. 132*, 2461.
Holland, M.G. (1966), *Physics of III–V Compounds* (eds. R.K. Willardson and A.C. Beer) Vol. 2, Acad. Press, New York.
Holland, M.G. (1971), *Phys. Rev. B3*, 3575.
Ioffe, A.F. (1956), *Can. J. Phys. 34*, 1342.
Jensen, H.H., Smith, H. and Wilkins, W. (1959), *Phys. Rev. 185*, 323.
Julian, C.L. (1965), *Phys. Rev. 137*, A128.
Keyes, R.W. (1959), *Phys. Rev. 115*, 564.
Klemens, P.G. (1951), *Proc. Roy. Soc. A208*, 108.
Klemens, P.G. (1955), *Phys. Rev. A68*, 1113.
Klemens, P.G. (1958), *Solid St. Phys.* (eds. F. Seitz and D. Turnbull), Acad. Press, New York, Vol 7.
Klemens, P.G. (1960), *Phys. Rev. 119*, 507.
Klemens, P.G. (1969), *Thermal Conductivity* (ed. R.P. Tye) Acad. Press London, Vol. 1.
Klemens, P.G. and Ecsedy, D.J. (1976), *Proc. 2nd Int. Conf. on Phonon Scatt. in Solids*, Nottingham, 1975 (ed. R.J. Challis *et al.*), Plenum Press, London, p. 367.
Kosarev, V.V., Tamarin, P.V. and Shalyt, S.S. (1971), *Phys. Stat. Solidi (b) 44*, 525.
Lawson, A.W. (1957), *J. Phys. Chem. Solids 3*, 155.
Leibfried, G. and Schlomann, E. (1954), *Nach. Akad. Wiss. Gottingen, Mat. Phys. Klasse 4*, 71.
Maradudin, A.A. (1964), *J. Amer. Chem. Soc. 86*, 3405.
Meddins, H.R. and Parrott, J.E. (1976), *J Phys. C: Solid St Phys.* 9 1263.
Men', A.A and Sergeev, O.A. (1973), *High temperatures-High Pressures 5*, 19.
Mizushima, S. (1954), *J. Phys. Soc. Japan 9*, 546.
Nath. P. and Chopra, K.L. (1974), *Phys. Rev. B10*, 3412.
Nettleton, R.E. (1963), *Phys. Rev. 132*, 2032.

Oliver, D.W. and Slack, G.A. (1966), *J. Appl. Phys. 37*, 1542.
Parrott, J.E. (1963), *Proc. Phys. Soc. 81*, 726.
Parrott, J.E. (1969), *J. Phys. C. 2*, 147.
Parrott, J.E. (1971), *Phys. Stat. Solidi (b) 48*, K159.
Parrott, J.E. and Stuckes, A.D. (1975), *Thermal Conductivity of Solids*, Pion Limited, London.
Pavlov, S.T. and Eshpulatov, B.E. (1977), *Sov. Phys.–Solid State 18*, 683.
Peierls, R.E. (1929), *Annln. Phys. 3*, 1055.
Pomeranchuk, I. (1941), *Phys. Rev. 60*, 820.
Pomeranchuk, I. (1941), *J. Phys. USSR 4*, 259.
Ranninger, J. (1965), *Phys. Rev. 140*, A2031.
Roufosse, M. and Klemens, P.G. (1973), *Phys. Rev. B7*, 5379.
Roufosse, M. and Klemens, P.G. (1974), *J. Geophys. Res. 79*, 703.
Savvides, N. and Goldsmid, H.J. (1973), *J. Phys. C: Solid St. Phys. 6*, 1701.
Savvides, N. and Goldsmid, H.J. (1974), *Phys. Stat. Solidi (b) 63*, K89.
Schieve, W.C. and Peterson, R.L. (1962), *Phys. Rev. 126*, 1456.
Slack, G.A. (1972), *Proc. Int. Conf. Phonon Scatt. in Solids* (ed. H.J. Albany), Centre d'Etudes Nucleaires de Saclay, p 24.
Slack, G.A. (1973), *J. Phys. Chem. Solids 34*, 321.
Slack, G.A. (1979), *Solid St. Physics* (ed. Ehrenreich *et al.*), Academic Press, New York, Vol 34.
Soma, T. (1977), *J. Phys. Soc. Japan 42*, 1491.
Srivastava, G.P. (1975), *Phys. Stat. Solidi (b) 68*, 213.
Srivastava, G.P. (1976), *J. Phys. C: Solid St. Phys. 9*, 3037.
Srivastava, G.P. (1981), *J. Phys. Colloq (France) 42*, C6—149.
Srivastava, G.P. and Hamilton, R.A.H. (1978), *Physics Reports 38C*, 3.
Steigmeier, E.F. and Abeles, B. (1964), *Phys. Rev. 136*, A1149.
Steigmeier, E.F. (1969), *Thermal Conductivity* (ed. R.P. Tye), Acad. Press, London. Vol 2, p. 203.
Sussman, J.A. and Thellung, A. (1963), *Proc. Phys. Soc. 81*, 11 2.
Verma, G.S., Bhandari, C.M. and Joshi, Y.P. (1971), *Phys. Rev. B3*, 3574.
Wagini, H. (1966), *Z. Naturforsch A21*, 2096.
White. G.K. and Woods, S.B. (1958), *Phil. Mag. 3*, 785.
Yates, B. (1972), *Thermal Expansion*, Plenum Press, New York.
Ziman, J.M. (1956), *Can. J. Phys. 34*, 1256.
Ziman, J.M. (1960), *Electrons and Phonons*, Clarendon Press, Oxford.

Chapter VII

Semiconducting Materials—Analysis of Experimental Data (I)

7.1 Introduction

In the previous chapter, various theoretical expressions employed in the computation of the lattice thermal conductivity of non-metallic solids were derived and their applicability in analysing experimental data discussed. The two approaches—the variational method and the relaxation-time method—have been used in the analysis of the experimental data of a large number of solids with varying degree of success. In spite of the difficulties in obtaining phonon relaxation times for various scattering mechanisms, the relaxation-time method has been successfully employed in the study of the thermal conductivity behaviour of solids. Chapters VII and VIII describe the important experimental and theoretical results for various semiconductors.

At low carrier densities, the thermal conductivity of semiconductors can readily be described in terms of phonon conduction where the phonon mean-free-path is limited mainly by other phonons, various impurities and imperfections and by crystal boundaries. The difficulties encountered in obtaining suitable relaxation times over extended ranges of temperature must, of course, be considered. In doped semiconductors where there is an appreciable concentration of free charge carriers or in the intrinsic conduction range, the electronic thermal conductivity (both polar and bipolar) must first be subtracted from the measured thermal conductivity. Any other contribution, such as that by photons, must also be taken into consideration before attempting a theoretical analysis of the lattice contribution.

Of the various formulations related to the relaxation-time approach, the Callaway formalism has been widely used in data analysis. A number of refinements of this method have been discussed (see Chap. VI).

In this chapter, we will discuss the thermal conductivity of elemental semiconductors and their alloys. The cases of silicon, germanium and their alloys have been described in detail as these materials have been studied intensively.

7.2 Group IV elements—germanium and silicon

This section and the following one describe results of various experimental measurements and the theoretical analysis of the lattice thermal conductivity of germanium, silicon and their alloys.

7.2.1 *Germanium*

The results of various measurements of the low temperature thermal conductivity of germanium are shown in Fig. 7.1 (a). The results agree in the temperature range 100–400 K, while below 100 K the differences appear to have been caused by differences in crystal purity.

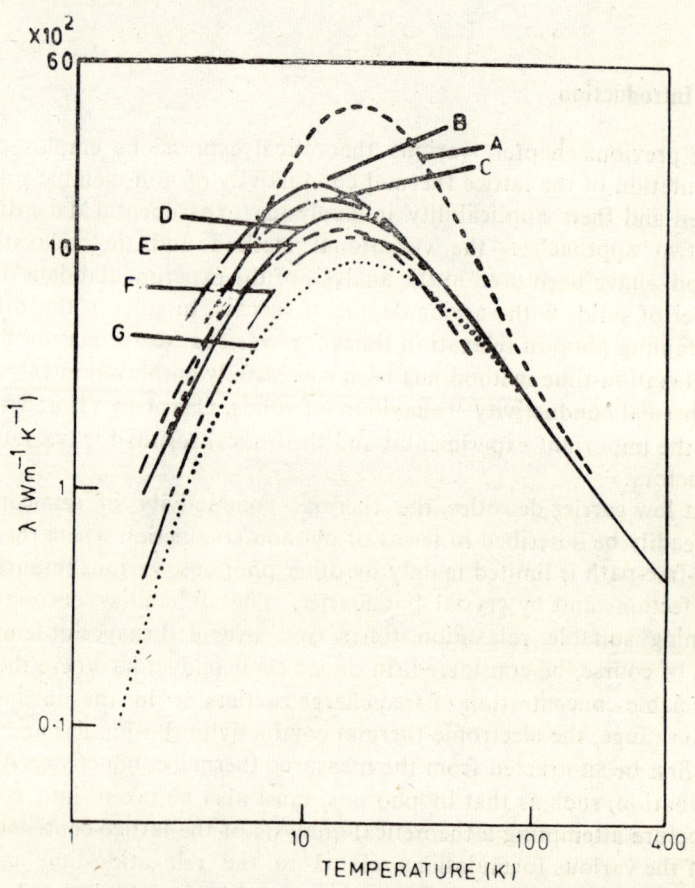

Fig. 7.1 (a) The low temperature thermal conductivity of germanium: Curve A, Geballe and Hull (1958) enriched Ge^{74}, carrier concentration 1.2×10^{19} m^{-3} (*n*-type); B, Glassbrenner and Slack (1964) 10^{20} m^{-3} (*p*-type); C, White and Woods (1955); D, Carruthers *et al* (1957) 10^{19} m^{-3} (*n*-type); E, Toxen (1961); F, Holland (1963): G, Goff and Pearlman (1965) 6.1×10^{21} m^{-3} (*n*-type).
(after Steigmeier, 1969).

SEMICONDUCTING MATERIALS—ANALYSIS OF EXPERIMENTAL DATA (I) 133

Analysis of the data can be carried out using the Callaway formalism described in the previous chapter. In order to explain the thermal conductivity over a wide range of temperatures (2–1000 K), a more refined form of the model was used by Holland (1963). He took into consideration the separate contributions from the transverse and the longitudinal phonon

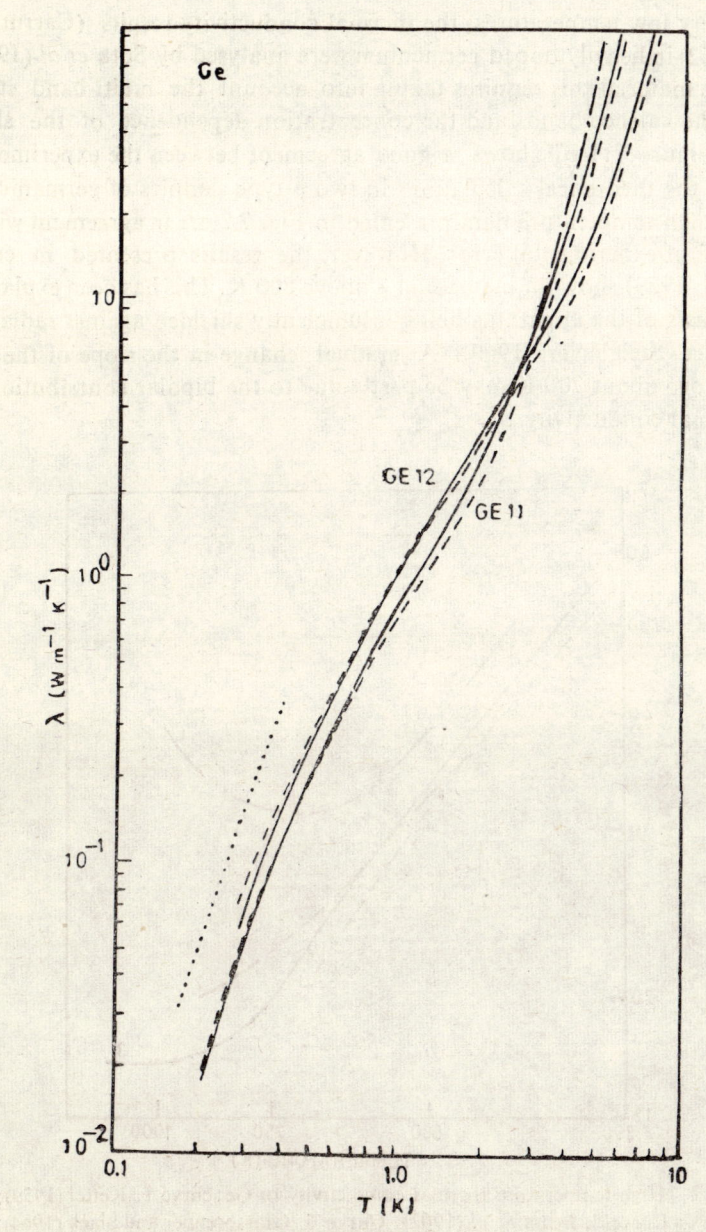

Fig. 7.1 (b) Very low temperature thermal conductivity of *p*-type germanium samples (after Sota *et al*, 1984).

modes together with the strong dispersion of the transverse acoustic (TA) branch; the correction term was neglected in his analysis. The TA branch was further subdivided into two parts having separate group velocities for phonons. Using this model, Holland obtained an excellent fit to the measured thermal conductivity over most of the temperature range from 2–1000 K.

At very low temperatures, the thermal conductivity results (Carruthers *et al*, 1962) in heavily doped germanium were analysed by Sota *et al* (1984). In *p*-type material this requires taking into account the multi-band structure of the valence bands and the concentration dependence of the shear terms. Figure 7.1(b) shows a good agreement between the experimental data and the theoretical calculations in two *p*-type samples of germanium.

The high temperature data presented in Fig. 7.2 are in agreement within the range of experimental error. However, the results presented in curve 1 (Kettel, 1959) show an increase in λ above 800 K. This has been explained on the basis of the apparatus being insufficiently shielded against radiation losses (see Steigmeier, 1969). A gradual change in the slope of the λ-T curve above about 700 K may be partly due to the bipolar contribution to the thermal conductivity.

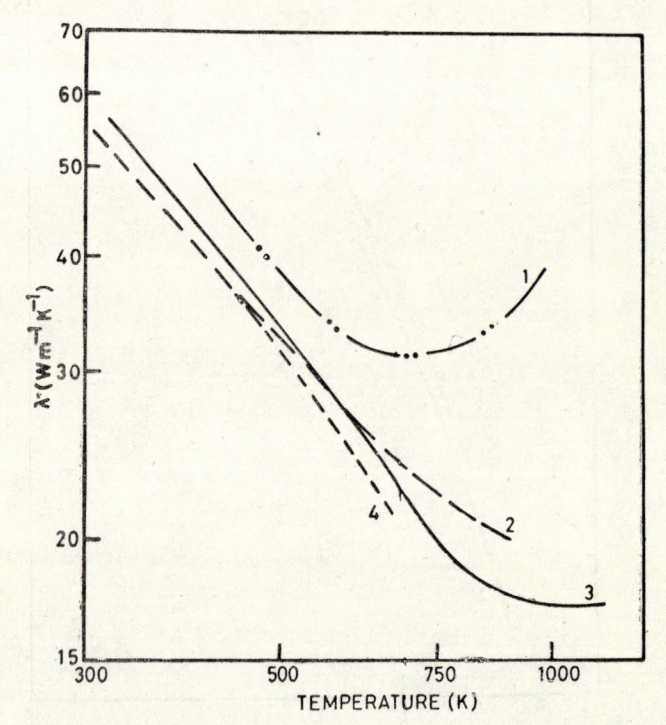

Fig. 7.2 High temperature thermal conductivity of Ge; curve 1, Kettel (1959); Curve 2, Beers *et al* (1962); Curve 3, Glassbrenner and Slack (1964); Curve 4, Stuckes (1960).
(after Steigmeier, 1969).

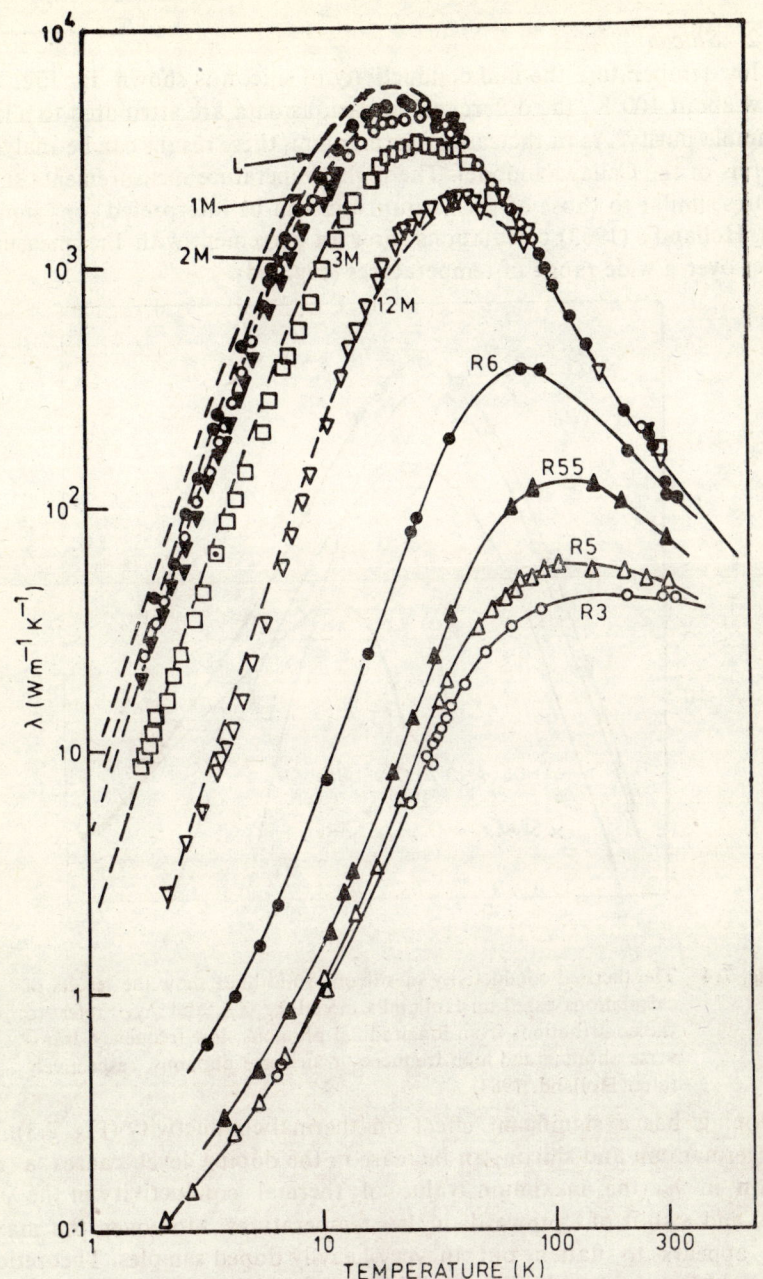

Fig. 7.3 Low temperature thermal conductivity of Si; 1M, 2M, 3M and 12M correspond to p-type material with carrier concentrations 10^{19} m^{-3}, 4.2×10^{20} m^{-3}, 4.0×10^{21} m^{-3} and 4.0×10^{22} m^{-3}, respectively (Holland and Neuringer 1962); ———— theoretical curves based on the Callaway model, curve L—no adjustable parameter, others-one adjustable parameter. R6 and R55 refer to n-type material with carrier concentrations 2.0×10^{25} m^{-3} and 1.7×10^{26} m^{-3}; R5 and R3 refer to p-type with 3×10^{26} m^{-3} and 5×10^{26} m^{-3} (Slack, 1964)(after Steigmeier 1969).

7.2.2 Silicon

The low-temperature thermal conductivity of silicon is shown in Fig. 7.3. Below about 100 K, the differences in various data are attributed to a lack of sample purity. As in the case of germanium, these results can be analysed in terms of the Callaway model. The high-temperature measurements show features similar to those of germanium and can be interpreted on similar lines. Holland's (1963) calculations show an agreement with the measured values over a wide range of temperatures (Fig. 7.4).

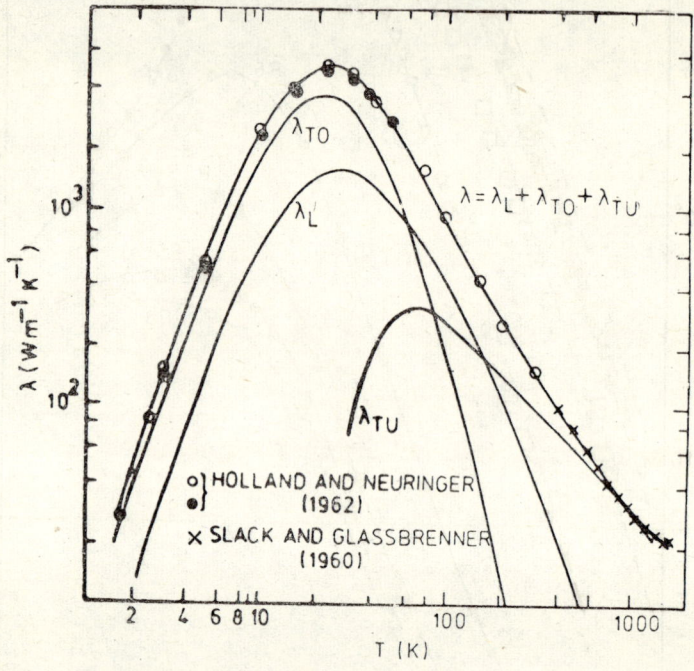

Fig. 7.4 The thermal conductivity of silicon. Solid lines show the results of calculations based on Holland's model. λ_L, λ_{Ta} and λ_{TU} refer to the contributions from longitudinal phonons, low frequency transverse phonons and high frequency transverse phonons, respectively (after Holland, 1963).

Doping has a significant effect on thermal conductivity (Fig. 7.3). In both germanium and silicon, an increase in the doping level causes a reduction in λ_m (the maximum value of thermal conductivity in the $\lambda-T$ curve) and a shift of λ_m towards higher temperatures. Moreover, the maximum appears to flatten out in very heavily doped samples. Theoretical curves obtained on the basis of the Callaway model show good agreement with experimental data. The curve marked L has no adjustable parameter while the other curves use one adjustable parameter for boundary scattering (Steigmeier, 1969).

The results of thermal conductivity measurements at very low temperatures were analysed by Sota *et al.* (1984) on lines similar to those reported for germanium. However, the agreement between the experimental and theoretical results is not as satisfactory as with germanium. This difference has been ascribed to the simple model used, which assumes a spherical approximation and neglects the anisotropy of the valence band.

The effect of doping on thermal conductivity can be understood in terms of the scattering of phonons by electrons bound to the impurity atoms and also by free carriers. These aspects of the phonon scattering and other related topics are discussed in Chap. V.

7.3 Alloys of silicon and germanium

The thermal conductivity of silicon-germanium alloys has been extensively studied over a range of compositions. It decreases considerably on alloying mainly because of point-defect scattering. The relaxation time for the scattering of phonons by point-defects depends upon a factor $f_i(1-M_i/\bar{M})^2$, where M_i is the mass of the ith type of impurity, $\bar{M}$ the mean atomic-mass and f_i the fractional concentration of this impurity (see Sec. 5.9). Silicon and germanium have widely differing atomic masses and, therefore, the point-defect scattering is relatively high. Consequently, the thermal conductivity of silicon-germanium alloys is almost an order of magnitude lower than that of the constituents (Fig. 7.5). For doped alloys there is a

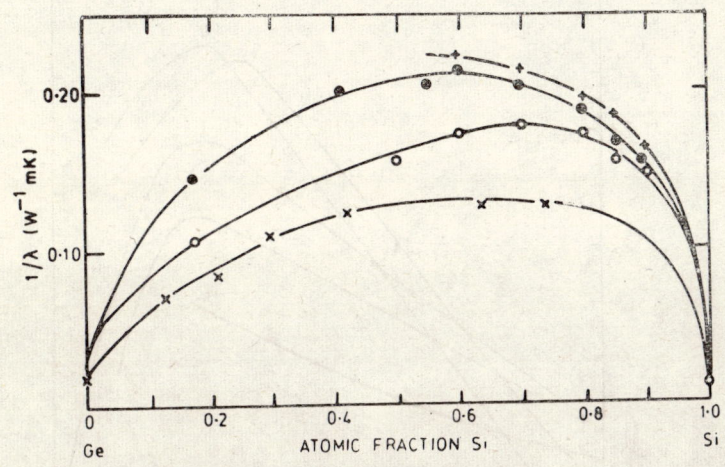

Fig. 7.5 The room temperature thermal resistivity of silicon-germanium alloys as a function of alloy composition : ×, undoped; o, 1.5×10^{26} m^{-3} (p-type); ●, 1.5×10^{26} m^{-3} (n-type) +, 1.5×10^{26} m^{-3} (n-type). Three doped samples are doped with B, As and P dopants, respectively. (after Dismukes *et al*, 1964).

further reduction in λ due to the phonon-electron scattering. In undoped alloys, the experimental data have been analysed in a satisfactory manner in terms of three-phonon and point-defect scattering (Abeles, 1963, and Parrott, 1963).

Dismukes *et al* (1964) measured the thermal conductivity of a large number of *n*- and *p*-type samples of Si-Ge alloys. A reasonably satisfactory analysis of the effect of doping on the lattice thermal conductivity was presented by Steigmeier and Abeles (1964). The relaxation time for the scattering of phonons by free electrons (or holes) was used and the deformation potential varied to obtain a best fit with the experimental data. This analysis was later extended by Bhandari and Verma (1965) by taking into account the separate contributions from the transverse and the longitudinal phonon branches. Erofeev *et al* (1965) rejected phonon-electron scattering on the grounds that an appreciable phonon-electron interaction should show up as a phonon-drag effect in the Seebeck coefficient which is not actually the case. However, the absence of a phonon-drag effect has been interpreted (Steigmeier, 1969) on the grounds that among the electron scattering mechanisms impurity and alloy scattering are dominant whereas for phonons the scattering by electrons is an important scattering mechanism at relatively high carrier concentrations. It appears that the absence of an appreciable phonon-drag requires a more careful investigation, as amongst the electron scattering processes the important mechanisms include scattering by acoustic phonons in a large number of semiconductors (see Chap. IV).

In Figs. 7.6(a) and (b) are displayed the results of the thermal conduc-

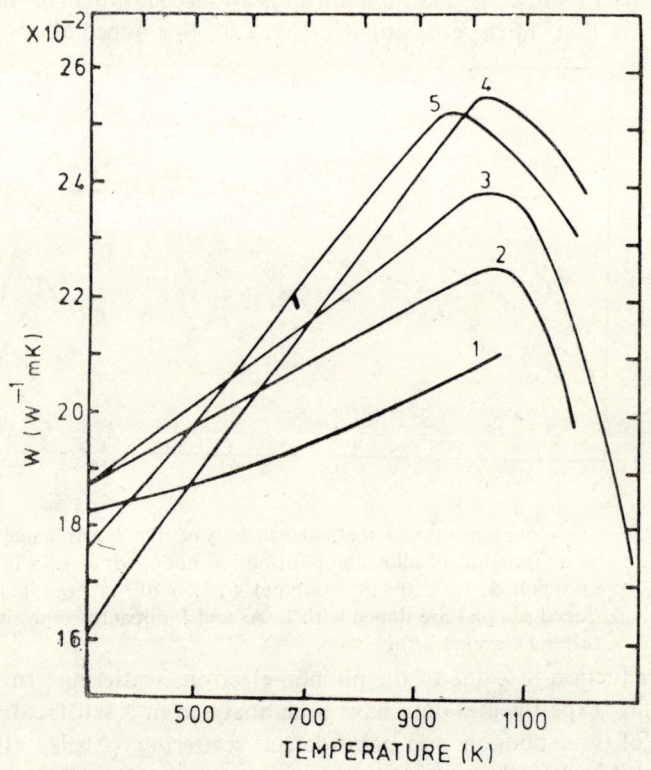

Fig. 7.6 (a) Thermal resistivity (W) of p-type $Si_{0.70}Ge_{0.30}$ alloys as function of temperature and carrier concentration; curve 1, 3.5×10^{26} m^{-3}, 2, 2.4×10^{26} m^{-3}; 3, 1.8×10^{26} m^{-3}; 4, 3.3×10^{25} m^{-3}; 5, 8.9×10^{25} m^{-3}. (after Dismukes *et al*, 1964).

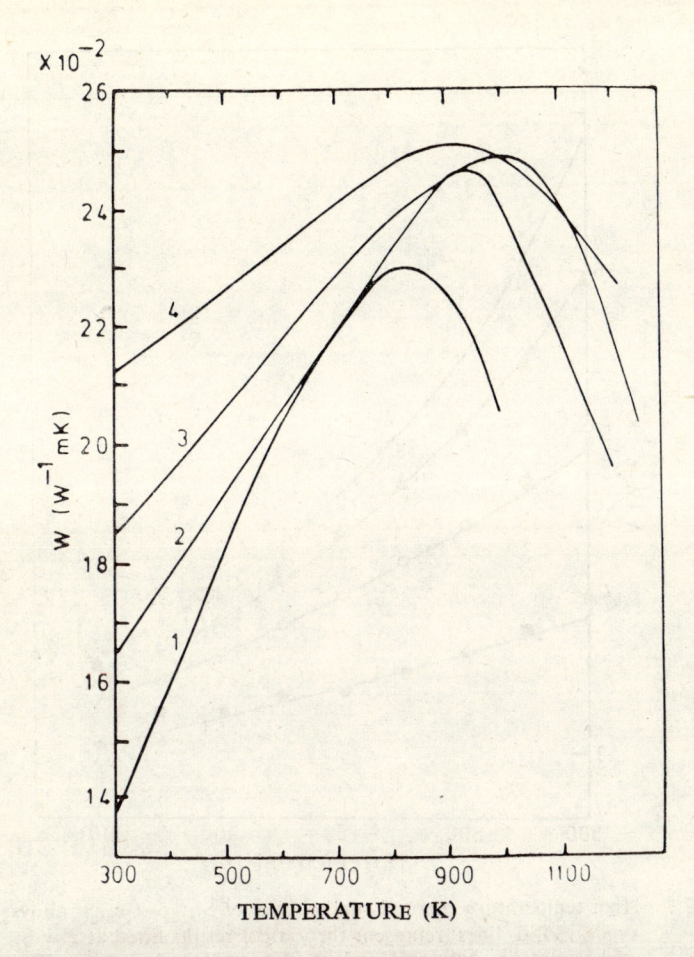

Fig. 7.6 (b) Thermal resistivity versus temperature in *n*-type Si-Ge alloys: curve 1, 2.2×10^{24} m^{-3}; curve 2, 2.3×10^{25} m^{-3}; curve 3, 6.7×10^{25} m^{-3}, curve 4, 1.5×10^{26} m^{-3}.
(after Dismukes *et al*, 1964).

tivity measurements on *p*- and *n*-type Si-Ge alloys as a function of temperature and carrier concentration (Dismukes *et al*, 1964). Figure 7.7 presents the results of theoretical calculations along with the experimental data (Steigmeier and Abeles, 1964). Gaur *et al.* (1966) presented an analysis of the data by calculating the contributions to thermal conductivity from the transverse and the longitudinal phonon branches.

7.4 Radiative heat transfer

The analysis of high-temperature data on thermal conductivity presents some difficulties. There are appreciable contributions to λ from electrons, holes and electron-hole pairs apart from the lattice contribution which is

140 THERMAL CONDUCTION IN SEMICONDUCTORS

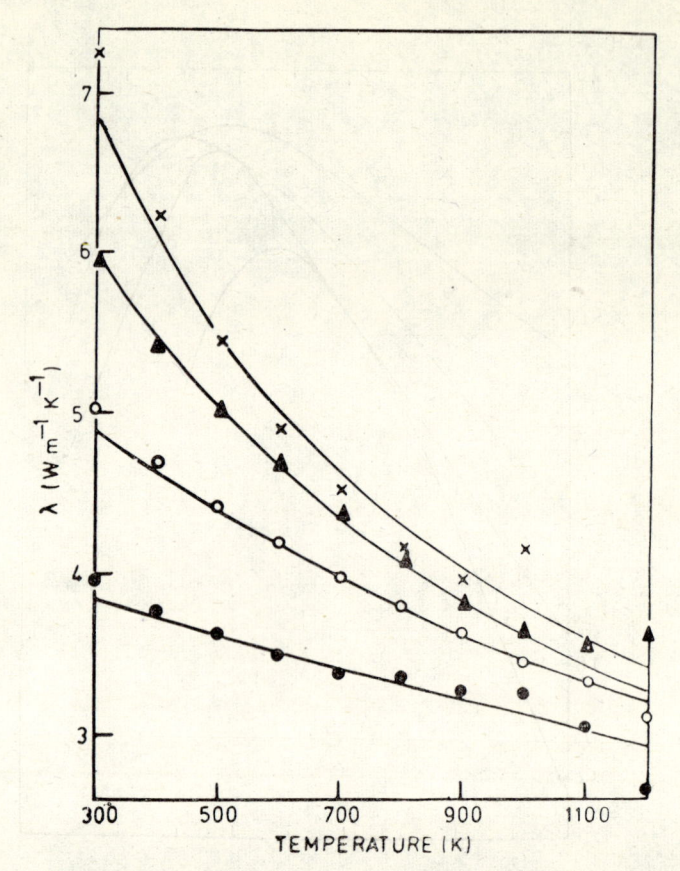

Fig. 7.7 High temperature thermal conductivity of $Si_{0.70}-Ge_{0.30}$ alloys (n-type). Solid lines represent theoretical results fitted at $T = \theta_D$ for each sample: x, 2.2×10^{24} m^{-3}; Δ, 2.3×10^{25} m^{-3}; o, 6.7×10^{25} m^{-3}; $\bullet$, 1.5×10^{26} m^{-3}.
(after Steigmeier and Abeles, 1964).

usually the most dominant contribution in non-metallic solids. Photon thermal conductivity may also become significant at these temperatures, and must be first of all subtracted from the measured thermal conductivity. The remaining ($\lambda_{meas} - \lambda_{photon}$) can then be analysed in terms of electronic and lattice contributions.

The photon thermal conductivity is given by Eq. 6.60. The temperature dependence of λ_{photon} arises due to the temperature variation of α^{-1} and the explicit T^3 term. In the extrinsic range of conduction, α^{-1} rises very slowly with temperature and, consequently, λ_{photon} rises rapidly with increasing temperature. With the onset of intrinsic conduction α^{-1} starts decreasing very rapidly with increase in temperature, which results in a decrease of the photon thermal conductivity. In a heavily doped semiconductor with a large concentration of carriers, α^{-1} becomes very small and λ_{photon} is almost negligible. These considerations point towards a maximum in the λ_{photon}-versus-temperature curve.

Beers et al (1962) made measurements on the thermal conductivity of silicon-germanium alloys and analysed the data by taking account of the photon thermal conductivity. Fig. 7.8 shows the effect of the photon contribution to the thermal conductivity of n-type Si-Ge alloys. The doped sample

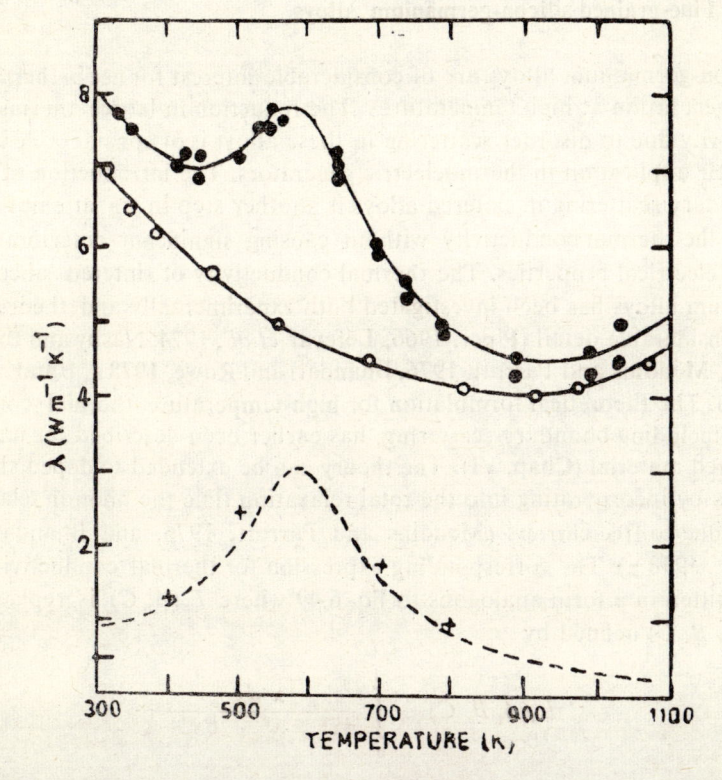

Fig. 7.8 The effect of photon contribution on the high temperature thermal conductivity of Ge-Si alloys (after Beers et al 1962). ●, near intrinsic; o, lightly doped; — — —, difference in λ values of the two measurements; +, λ_{photon} (calculated).

shows a smooth temperature variation while the one with a lower carrier concentration shows a peak. The difference between the two curves may arise due to photon and electronic contributions. Assuming that the difference arises mainly due to photon contribution, λ_{photon} may be obtained by subtracting one from the other. The resulting curve is in good agreement with the theoretically calculated photon thermal conductivity.

In pure silicon and germanium a photon contribution has not been observed, whereas it has been detected in gallium arsenide. In pure silicon the thermal conductivity is quite large and λ_{photon} is difficult to observe as it forms a small fraction of the total thermal conductivity. In germanium a smaller energy gap entails a very low value of α^{-1} and therefore λ_{photon} is small. A semiconductor with a relatively large gap and low thermal conductivity is more likely to show an appreciable photon thermal conductivity. The bipolar contribution λ_b and the photon contribution λ_{photon} may not be significant in the same range of temperature, as much higher temperatures are required for a significant bipolar contribution.

7.5 Fine-grained silicon-germanium alloys

Silicon-germanium alloys are of considerable interest for use in thermoelectric generation at high temperatures. The reduction in lattice thermal conductivity due to disorder scattering in these alloys is of considerable interest in their application in thermoelectric generators. The introduction of grain-boundary scattering in sintered alloys is another step in an attempt to reduce the thermal conductivity without causing significant deterioration in their electrical properties. The thermal conductivity of sintered silicon-germanium alloys has been investigated both experimentally and theoretically in considerable detail (Piper, 1966, Lefevar *et al.*, 1974, Nasby and Burgess, 1972, Meddins and Parrott, 1976, Bhandari and Rowe, 1978 a, b and Rowe, 1982). The theoretical formulation for high-temperature thermal conductivity, including boundary scattering, has earlier been described for undoped sintered material (Chap. VI). The theory can be extended to doped sintered alloys by incorporating into the total relaxation time the phonon relaxation time due to free carriers (Meddins and Parrott, 1976, and Bhandari and Rowe, 1978 a). The corresponding expression for thermal conductivity can be written in a form analogous to Eq. 6.49 where $L_n(A, C)$ is replaced by $L_n(A, B, C)$ defined by

$$L_n(A, B, C) = \int_0^1 \frac{x^n dx}{Ax^4 + x^2 + Bx + C}$$

The additional relaxation time due to the scattering by free carriers is taken to be proportional to phonon frequency ω (Parrott 1979; also see section 5.10).

A refined version of these expressions which takes account of the separate contributions from the different polarisation branches was given by Bhandari and Rowe (1978 a) and the calculated values of thermal conductivity were found to be in agreement with the observed decrease in the thermal conductivity of thin films of silicon (Savvides and Goldsmid, 1973, and 1974). The agreement was, however, found to be poor for doped sintered alloys of Si and Ge. Figure 7.9 gives a plot of calculated lattice thermal conductivity as a function of carrier concentration and grain-size in p-type $Si_{0.70}Ge_{0.30}$ alloys (Bhandari and Rowe, 1978 b).

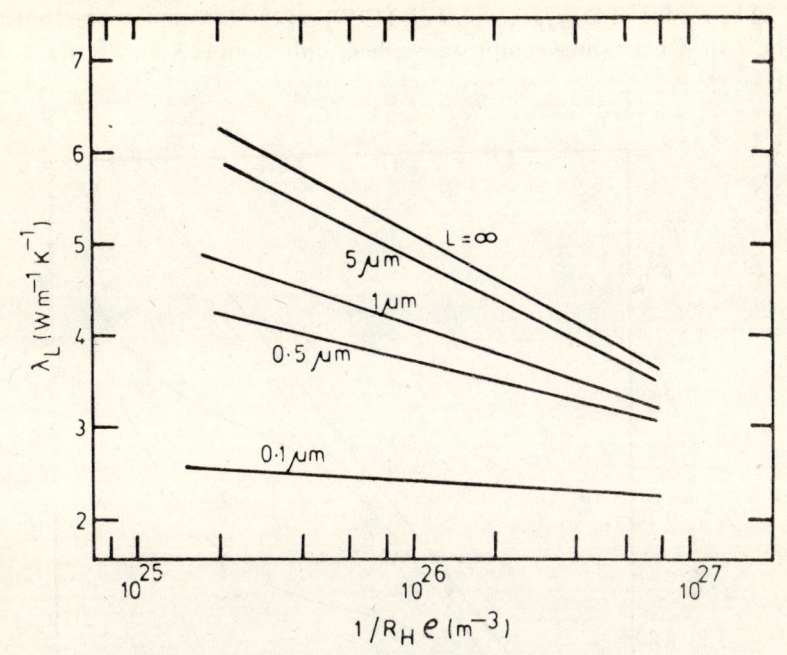

Fig. 7.9 Lattice thermal conductivity (calculated) of $Si_{0.70}$ $Ge_{0.30}$ alloys as a function of carrier concentration and grain size.
(after Bhandari and Rowe, 1978 b)

7.6 Other elemental semiconductors

7.6.1 *Selenium and tellurium*

White *et al* (1957) measured the thermal conductivity of crystalline and amorphous selenium. The thermal conductivity of amorphous selenium is about two orders of magnitude lower than that of the crystalline material and decreases steadily as the temperature is reduced (see Fig. 9.1). Abdullaev *et al.* (1966) reported high temperature measurements in selenium.

Addition of halogen impurities leads to a reduction in thermal conductivity. Measurements on both amorphous and crystalline material containing chlorine, bromine and iodine impurities have been reported by Aliev and Abdullaev (1957), Abdullaev and Bashshaliev (1957) and Abdullaev and Aliev (1957).

Fischer *et al.* (1957) reported thermal conductivity measurements on a number of tellurium samples down to around 2K. Near the conductivity maximum λ is mainly limited by isotope scattering as there are a large number of isotopes present in natural tellurium. Below 5K, thermal conductivity is limited by boundary scattering. High-temperature measurements were reported by Amirkhanova *et al.* (1959), Devyatkova *et al.* (1959), Smirnov and Shadrichev (1963) and Sutter and Galo (1965). A large anisotropy was observed in thermal conductivity measured along the *c*-axis and in a direction perpendicular to it.

The results of Devyatkova *et al.* (1959) on three samples are shown in Fig. 7.10. Of the three samples described, only sample 3 shows the normal

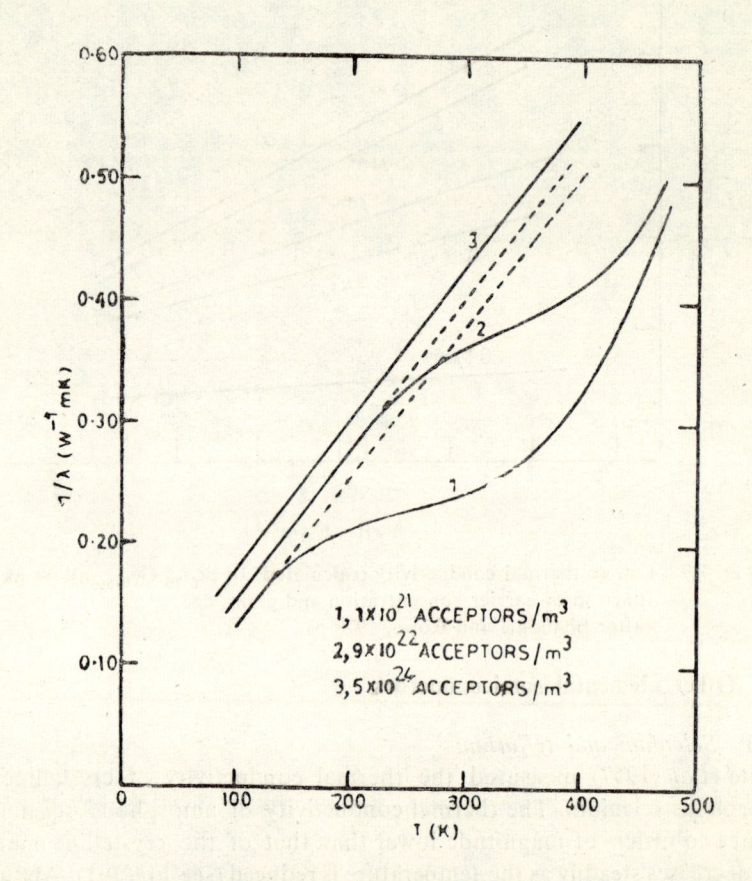

Fig. 7.10 The thermal resistivity of tellurium as a function of temperature (after Devyatkova *et al*, 1959).

behaviour expected when phonon-phonon scattering is the dominant phonon scattering mechanism. The thermal resistivity of sample 1 falls below the expected value above around 130 K and that of sample 2 around 200 K. Devyatkova and coworkers attributed this to the presence of photon contribution to thermal conductivity. This contribution is expected to be absent in samples 3 due to a high absorption of photons by free carriers. Good agreement was reported between the theoretical results obtained on the basis of Eq. (6.60) and the experimental data.

Figure 7.11 shows the variation of the thermal conductivity of Se and Te with temperature.

SEMICONDUCTING MATERIALS-ANALYSIS OF EXPERIMENTAL DATA (I) 145

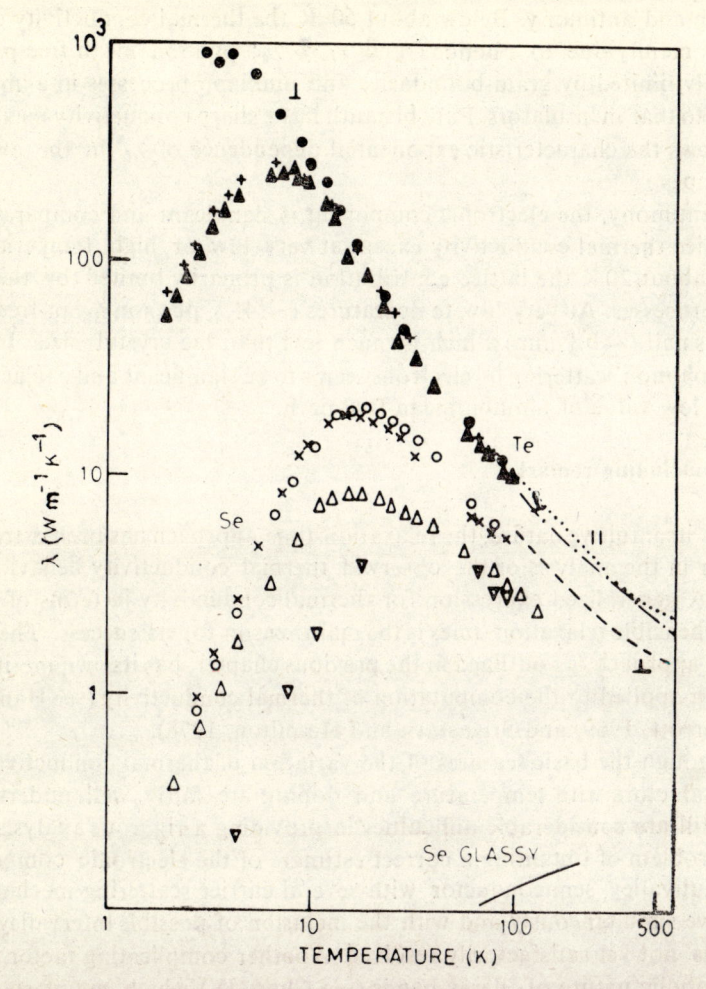

Fig. 7.11 The thermal conductivity of selenium and tellurium: o, x, △, ▽, polycrystalline Se (White *et al*, 1958) ●, Te single crystal ⊥ to *c*-axis (Fischer *et al*, 1957); ▲, +, polycrystalline Te (Fischer *et al*, 1957). ......Te single crystal ∥ to *c*-axis (Amirkhanova *et al*, 1959), —.—.—. Te single crystal ∥ to *c*-axis (Smirnov and Shadrichev, 1963), ——— Te single crystal ⊥ to *c*-axis (Amirkhanova *et al*, 1959) (after Steigmeier, 1969).

7.6.2 Other materials

It is worthwhile to give a brief description of the thermal conductivity of bismuth and antimony. Although not included in the list of semiconductors, these elements display some of the features of degenerate semiconductors and their lattice thermal conductivity behaviour is unlike that of metals.

Shalyt (1944), Rosenberg (1955), and White and Woods (1958) have reported measurements of the low-temperature thermal conductivity of bismuth and antimony. Below about 50 K the thermal conductivity of bismuth is mainly due to phonons and $\lambda_L \gg \lambda_e$; phonon mean free-path is primarily limited by grain-boundaries and umklapp processes in a manner similar to that in insulators. Pure bismuth has a sharp conductivity maximum and shows the characteristic exponential dependence of λ_L in the absence of isotopes.

In antimony, the electronic component is significant and comparable to the lattice thermal conductivity except at very low or high temperatures. Above about 20 K the lattice contribution is primarily limited by the umklapp processes. At very low temperatures ($\sim$ 4K), phonon mean-free-path is very small ($\sim$ 0.1 mm) which is much less than the crystal size. In this region phonon scattering by electrons seems to be significant and can account for the low value of phonon mean-free-path.

7.7 Concluding remarks

Despite its intuitive nature, the relaxation-time approach has been extremely popular in the analysis of the observed thermal conductivity behaviour of solids. A generalized expression for thermal conductivity in terms of easily comprehensible relaxation times is the main reason for its success. The variational approach, as outlined in the previous chapter, has its own merits and has been applied to the computation of thermal conductivity (see Hamilton and Parrott, 1969, and Srivastava and Hamilton, 1978).

Although the basic features of the variation of thermal conductivity of semiconductors with temperature and doping are fairly well understood, there still are considerable difficulties in providing a rigorous analysis. The basic problem of obtaining a correct estimate of the electronic component in a mutivalley semiconductor with several carrier scattering mechanisms operative simultaneously and with the inclusion of possible intervalley scattering is not yet satisfactorily resolved. Another complicating factor is the nonparabolic nature of energy bands (see Chap. IV) which manifests itself in heavily doped materials specially in narrow-band-gap semiconductors.

The thermal conductivity of heavily doped semiconductors at very low temperatures (Sota *et al*, 1984), the effect of thermal expansion (Slack, 1979, also Chap. VI of this book), the part played by optical phonons in heat transport (Slack, 1979) and the need for a proper generalization of the thermal conductivity expression (Parrott, 1971) are but some of the topics which have to be carefully examined in a quantitative analysis of the observed data.

References

Abeles, B. (1959), *J. Phys. Chem. Solids 8*, 340.
Abeles, B. (1963), *Phys. Rev. 131*, 1906.
Abdullaev, G.B. and Aliev, M.I. (1957), *Dokl. Akad. Nauk. SSSR 114*, 995.

Abdullaev, G.B. and Bashshaliev, A.A. (1957), *Zh. Tekh. Fiz. 27*, 1971.
Abdullaev, G.B., Aliev, G.M. and Safaraliev, G.I. (1966), *Phys. Stat. Solidi 17*, 777.
Aliev, G.M. and Abdullaev, G.B. (1957), *Dokl. Akad Nauk. SSSR 116*, 598.
Amirkhanova, K.I., Bagduev, G.B and Kazhaev, M.A. (1959), *Sov. Phys. Dokl. 4*, 115.
Beers, D.S., Cody, G.D. and Abeles, B. (1962), *Proc. of the International Conf. on the Phys. of Semicond.*, Exeter, p. 41, The Instt. of Physics, London.
Bhandari, C.M. and Verma, G.S (1965), *Phys. Rev. 140*, A2101.
Bhandari, C.M. and Rowe, D.M. (1978 a), *J. Phys. C: Solid St. Phys. 11*, 1787.
Bhandari, C.M. and Rowe, D.M. (1978 b), *Proc. of 2nd International Conf. on Thermoelectric Energy Conversion*, Univ. of Texas at Arlington, p. 32.
Carruthers, J.A., Cochran, J.F. and Mendelssohn, K. (1962), *Cryogenics 2*, 160.
Carruthers, J.A., Geballe, T.H., Rosenberg, H.M. and Ziman, J.M. (1957), *Proc. Roy. Soc. A238*, 502.
Devyatkova, E.D., Moizhes, B.Ya and Smirnov, I.A. (1959), *Sov. Phys.-Solid St. 1*, 555.
Dismukes, J.P. and Ekstrom, L. (1965), *Trans. Am. Instt. Min Metal Engrs. 233*, 672.
Dismukes, J.P., Ekstrom, L., Steigmeier, E.F., Kudman, I. and Beers, D.S. (1964), *J. Appl. Phys. 45*, 2899.
Erofeev, R.S., Iordanishvili, E.K. and Petrov, A.V. (1965), *Sov. Phys.-Solid St. 7*, 2470.
Fischer, G., White, G K. and Woods, S.B. (1957), *Phys. Rev. 106*, 490.
Gaur, N.K.S., Bhandari, C.M. and Verma, G.S. (1966), *Phys. Rev. 144*, 628.
Geballe, T.H. and Hull, G.W. (1958), *Phys. Rev. 110*, 773
Goff, J.P. and Pearlman, N. (1965), *Phys. Rev. 140*, A2151,
Glassbrenner, C.G. and Slack, G.A. (1964), *Phys. Rev. 134*, A1058.
Hamilton, R.A.H. and Parrott, J.E. (1969), *Phys. Rev. 178*, 1284; also *Phys. Letters A29*, 556 (1969).
Holland, M.G. (1963), *Phys. Rev. 132*, 2461.
Holland, M.G. and Neuringer, L.J. (1962), *Proc. of International Conf. on Physics of Semicond.*, Exter, Instt. of Physics, London.
Kettel, F. (1959), *J. Phys. Chem. Solids 10*, 52.
Lefevar, R.A., McVay, G.L. and Baughman, R.J. (1974), Prep. of Hot-pressed Si-Ge Ingots, Part III—Vacuum hot pressing, *Mat. Res. Bull. 9*, p. 843.
Meddins, H.R. and Parrott, J.E. (1976), *J. Phys, C: Solid St. Phys. 9*, 1263.
Nasby, R.D. and Burgess, E.L. (1972), *Proc. of 7 th IECEC*, San Diego, California, p. 130.
Parrott, J.E. (1963), *Proc. Phys. Soc. 81*, 726.
Parrott, J.E. (1971), *Phys. Stat. Solidi (b) 48*, K 159.
Parrott, J.E. (1979), *Rev. Int. Hautes Temper. Refract. Fr., 16*, 393.
Piper, M. (1966), *Proc. UKAEA—ENEA Int. Symp. on Indus. Appl. for Isotopic Powered Generators*. Harwell, UKAEA, p. 349.
Rosenberg, H.M. (1955), *Phil. Trans. 247*, 441.
Rowe, D.M. (1982), *Proc. 4th Int. Conf. on Thermoelectric Energy Conv.*, Univ. of Texas at Arlington. March 1982.
Savvides, N. and Goldsmid, H J (1973), *J. Phys. C: Solid St. Phys. 6*, 1701.
Savvides, N. and Goldsmid H.J. (1974), *Phys. Stat. Solidi (b) 63*, K 89.
Shalyt, S. (1944), *J. Phys. USSR 8*, 315.

Slack, G.A. (1964), *J. Appl. Phys. 35*, 3460.

Slack, G.A. (1979), *Solid State Physics* (ed. H. Ehrenreich *et al.*), Academic Press, Vol. 34, p. 1.

Slack, G.A. and Glassbrenner, C.J. (1960), *Phys. Rev. 120*, 782.

Smirnov, I.A. and Shadrichev, E.V. (1963), *Sov. Phys.—Solid St. 4*, 1435.

Sota, T. and Suzuki, K. (1984), *J. Phys. C: Solid St. Phys. 17*, 2661.

Sota, T., Suzuki, K. and Fortier, D. (1984), *J. Phys. C: Solid St. Phys. 17*, 5935.

Srivastava, G.P. and Hamilton, R.A.H. (1978), *Phys. Reports 38 C*, 3.

Steigmeier, E.F. (1969), *Thermal Conductivity* (ed. R.P. Tye), Academic Press, London, Vol. 2.

Steigmeir, E.F. and Abeles, B. (1964), *Phys. Rev. 136*, A 1149.

Stuckes, A.D. (1960), *Phil. Mag. 5*, 84.

Sutter, P.H. and Galo, G.F. (1965) Bull. Amer. Phys. Soc. 10, 126.

Toxen, A.M. (1961), *Phys. Rev. 122*, 450.

White, G.K. and Woods, S.B. (1955), *Can. J. Phys. 33*, 58.

White, G.K. and Woods, S.B. (1958), *Phil. Mag. 3*, 342.

White, G.K., Woods, S.B. and Elford, M.T. (1958), *Phys. Rev. 112*, 111.

Ziman, J.M. (1956), *Phil. Mag. 1*, 191.

Ziman, J.M. (1957), *Phil. Mag. 2*, 292.

Chapter VIII

Semiconducting Materials—Analysis of Experimental Data (II)

8.1 Introduction

The analysis of thermal conductivity data of compound semiconductors follows the same pattern as described for silicon and germanium. The III-V compounds are amongst the most thoroughly investigated materials apart from silicon and germanium. These compounds can be obtained in a highly pure form which makes them suitable for the analysis of thermal conductivity data. They exhibit a wide range of lattice and electronic properties which influence their thermal conductivity behaviour. Detailed information on various useful physical properties, such as energy band gap, sound velocity, Debye temperature and impurity ionization energy, is available on these materials and this is of considerable help in the analysis of observed thermal conductivity. Other groups of semiconductors described in this chapter are: lead chalcogenides, II-VI, II-IV and V-VI compounds.

Holland (1966) and Steigmeier (1969) have reviewed the important experimental and theoretical aspects of the thermal conductivity of semiconductors. This chapter presents a relatively detailed account of the thermal conductivity of indium arsenide, indium antimonide, gallium arsenide and gallium antimonide (III-V group) and lead telluride (IV-VI group). The analysis of the thermal-conductivity data of other materials follows a similar pattern. The case of highly anisotropic (V-VI) group of compounds requires a special mention and these details are taken up in Chap. X.

8.2 Thermal conductivity of III-V compounds

8.2.1 *Indium antimonide*

For temperatures above the Debye temperature, the results of thermal conductivity measurements are shown in Fig. 8.1. Not all high-temperature data are in agreement and some show an increase of thermal conductivity with increasing temperature. It was realized (Steigmeier, 1969) that this

150 THERMAL CONDUCTION IN SEMICONDUCTORS

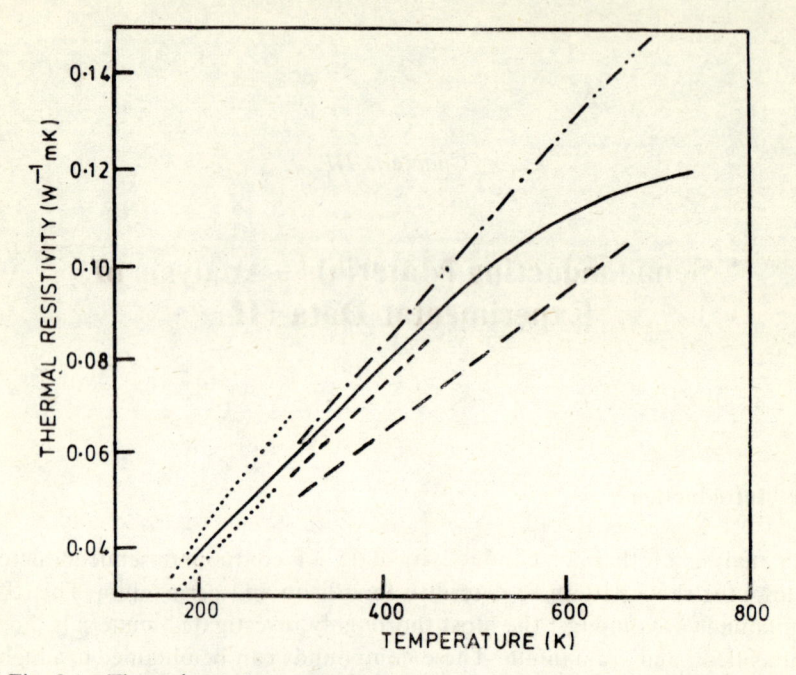

Fig. 8.1 Thermal resistivity of InSb at high temperature———— Busch and Steigmeier (1961) and Busch et al. (1959), carrier concentrations 1.6×10^{21} m^{-3} (p-type), 3.3×10^{21} m^{-3} (p-type) and 1.2×10^{22} m^{-3} (n-type); ......Amirkhanova and Bashirov (1960) 2×10^{22} m^{-3} (n-and p-types); ...Holland (1964a) 7×10^{19} m^{-3} (n-type); —.—.—. Wagini (1964) 3.5×10^{21} m^{-3} (n-type), 2.4×10^{22} m^{-3} (n-type), 5.7×10^{23} m^{-3} (p-type); —..—.. Wagini (1964) polycrystal-line; ————Wagini (1964) 8×10^{23} m^{-3} (n-type); —— Stuckes (1957) 2×10^{24} m^{-3} (n-type) (after Steigmeier, 1969).

apparent increase may be due to radiation losses during measurements (Busch and Schneider, 1954, and Weiss, 1958). Other independent measurements were made by Goldsmid (1954), Ioffe (1956), Zhuze (1954), Kanai and Nii (1959), Bowers et al. (1959), Bettman and Schneider (1960), Timberlake et al. (1962) and Wagini (1964); the last of these disagreeing with other authors above 500 K (not all these results are shown in the figure).

The thermal resistivity of doped n-type indium antimonide was measured by Stuckes (1957) and showed a somewhat lower value as compared to that of the undoped material. On the other hand, the p-type material showed a slightly higher thermal resistivity. The differences can be qualitatively explained on the basis of larger mobility of electrons as compared to that of holes; this gives rise to a larger electronic thermal conductivity in the doped (n-type) material. This appears to compensate for the decrease in the lattice thermal conductivity due to the scattering of phonons by electrons.

For temperatures well below the Debye temperature, the results of mea-

SEMICONDUCTING MATERIALS—ANALYSIS OF EXPERIMENTAL DATA (II) 151

surements by Mielczarek and Frederikse (1959) and Holland (1964 a) are shown in Fig. 8.2. Bhandari and Verma (1965) presented an analysis of the

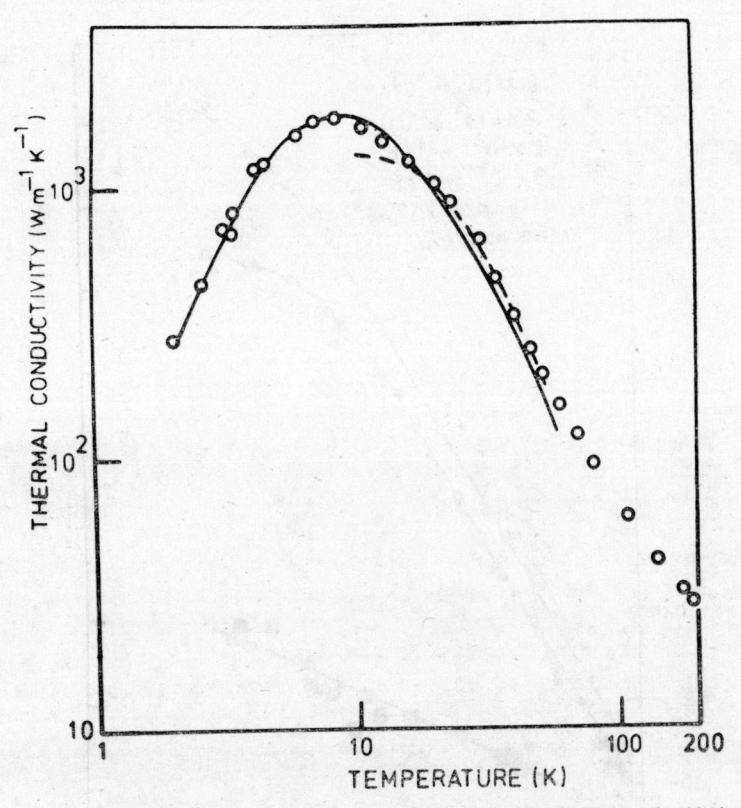

Fig. 8.2 Low-temperature thermal conductivity of InSb: o Holland (1964 a), 7×10^{19} m^{-3} (n-type); ———Holland (1964 a) theoretical; ———— Mielczarek and Frederikse (1959) 5×10^{24} m^{-3} (p-type) (after Steigmeier, 1969).

data by taking into account separate contributions to the thermal conductivity from transverse and longitudinal phonon branches. The effect of doping on low-temperature thermal conductivity has been investigated by Challis et al. (1964) between 1 and 5 K. The decrease in λ can be analysed in terms of phonon scattering by electrons in the bound states (Pyle, 1961, Keyes, 1961, Griffin and Carruthers, 1963) or by conduction electrons (Ziman, 1956, and 1957). The basic features are similar to those discussed in the following section for gallium antimonide.

8.2.2 Indium arsenide

Figure 8.3 shows the results of measurements on indium arsenide by Steigmeier and Kudman (1963) at high temperatures. For undoped samples (< 10^{22} m^{-3}) the data are in agreement with that of Stuckes (1957) but differ from that of Bowers et al. (1959). Some of this discrepancy may have resulted from the conversion of the measured thermal diffusivity into thermal conductivity.

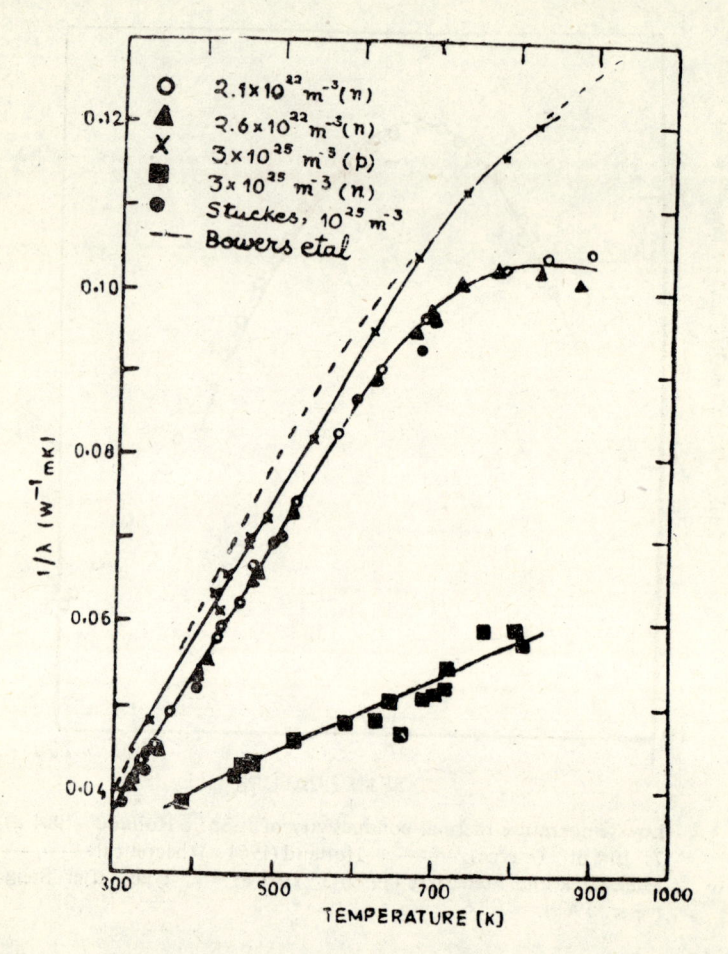

Fig. 8.3 High-temperature thermal resistivity of InAs (after Steigmeier and Kudman, 1963).

As in indium antimonide, the variation of thermal resistivity with increased doping shows different behaviour for *n*- and *p*-type impurities. The *n*-type impurity decreases the thermal resistivity whereas the *p*-type impurity tends to increase it. As described earlier in the case of indium antimonide this difference in the behaviour of *n*- and *p*-type materials can be explained in terms of a larger electronic contribution in *n*-type material. Both electrons and holes scatter phonons causing a reduction in the lattice thermal conductivity but a larger electronic thermal conductivity in *n*-type samples more than compensates for any decrease in λ_L due to an increase in carrier concentration (Steigmeier and Kudman, 1963).

8.2.3 *Gallium arsenide*

The high-temperature thermal conductivity of gallium arsenide is shown in Fig. 8.4. Sample 4 refers to the undoped crystal. An increase in the carrier

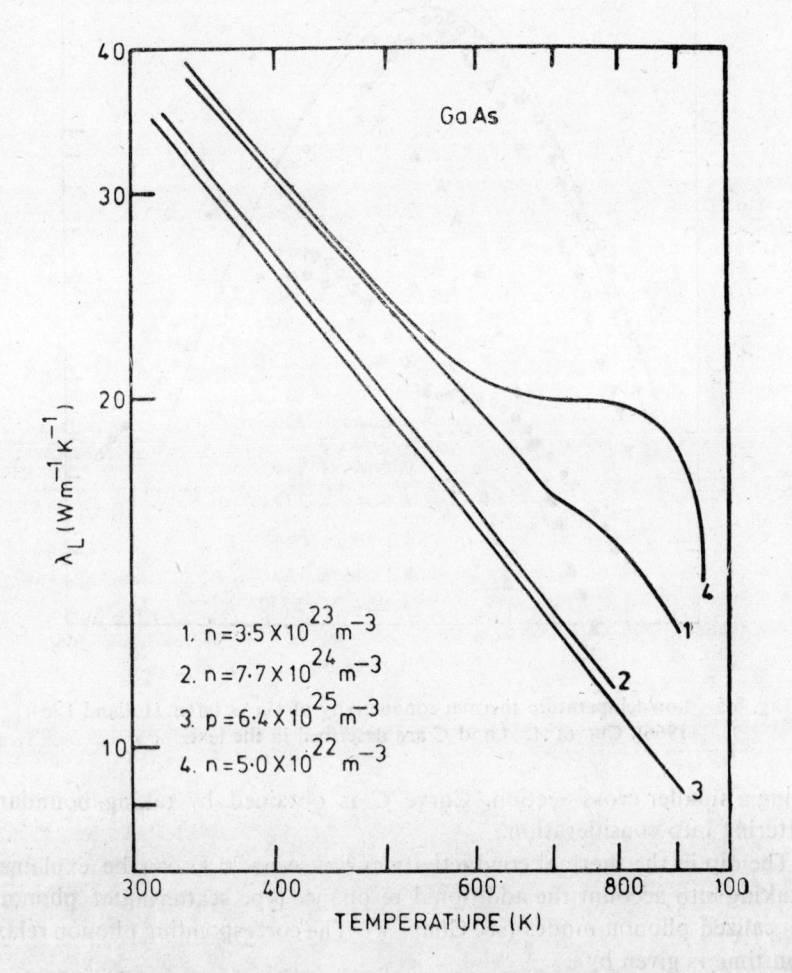

Fig. 8.4 High-temperature lattice thermal conductivity of GaAs (after Amith et al., 1965).

concentration causes a decrease in thermal conductivity, and this is ascribed to the scattering of phonons by electrons. The anomaly in the shape of curve 4 and to some extent in curve 1 has been attributed to the radiative transfer of heat. This type of behaviour had earlier been reported in other materials (Devyatkova *et al.*, 1959 and Beers *et al.*, 1962). Theoretical aspects of the radiative heat transfer have been discussed earlier in Chap. VI and VII.

The low-temperature measurements (Fig. 8.5) show a marked dip near 20 K (Holland, 1964 a, b). Curves *A*, *B* and *C* show the results of theoretical calculations. Samples 1 and 2 refer to the undoped material, sample 2

154   THERMAL CONDUCTION IN SEMICONDUCTORS

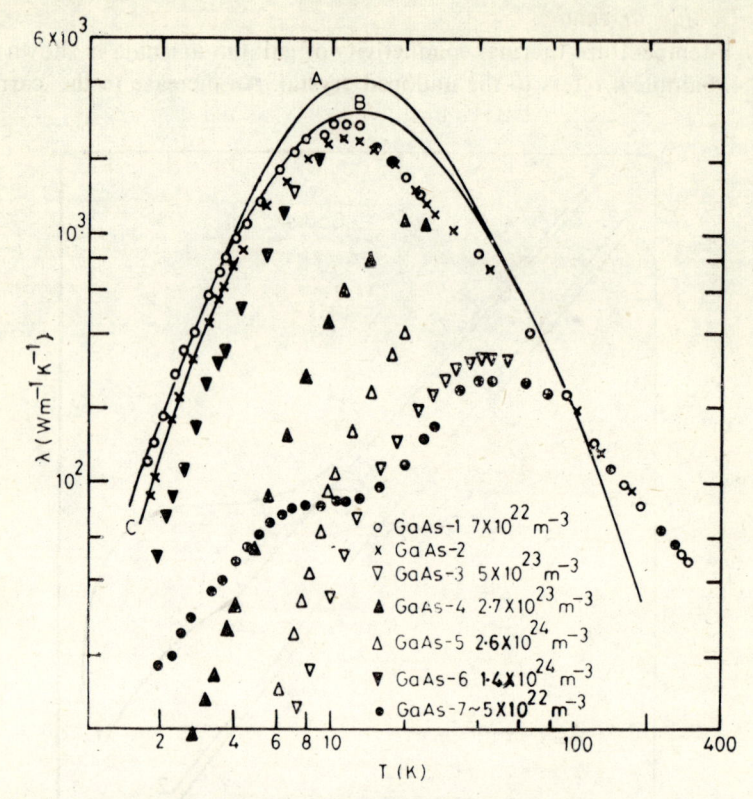

Fig. 8.5   Low-temperature thermal conductivity of GaAs (after Holland 1964, 1966). Curves A, B and C are described in the text.

having a smaller cross section. Curve C is obtained by taking boundary scattering into consideration.

The dip in the thermal conductivity curves near 20 K can be explained by taking into account the additional resonance type scattering of phonons by localized phonon modes (see Chap. V). The corresponding phonon relaxation time is given by

$$\tau^{-1}_{Resonance} = \frac{R\omega^2 T^2}{(\omega_0^2 - \omega^2)^2 + \left(\frac{\Lambda}{\pi}\right)^2 \omega_0^2 \omega^2}$$

where ω and ω_0 refer to the phonon and resonance frequencies respectively; R depends upon the concentration of impurity and Λ describes damping. The theoretically calculated thermal conductivity is found to be in agreement with the data (Gaur et al., 1966).

8.2.4 Gallium antimonide

Steigmeier and Kudman (1966) measured the high-temperature thermal conductivity of gallium antimonide. Other measurements by Guthrie and Whitefield (1966) and Amirkhanova and Magomedov (1966) are in good agreement with those of Steigmeier and Kudman. Abdullaev *et al* (1966) studied the effect of doping on thermal conductivity.

Figure 8.6 shows the measured thermal conductivity for various doping levels in both *n*- and *p*-type material at low temperatures (Holland, 1964).

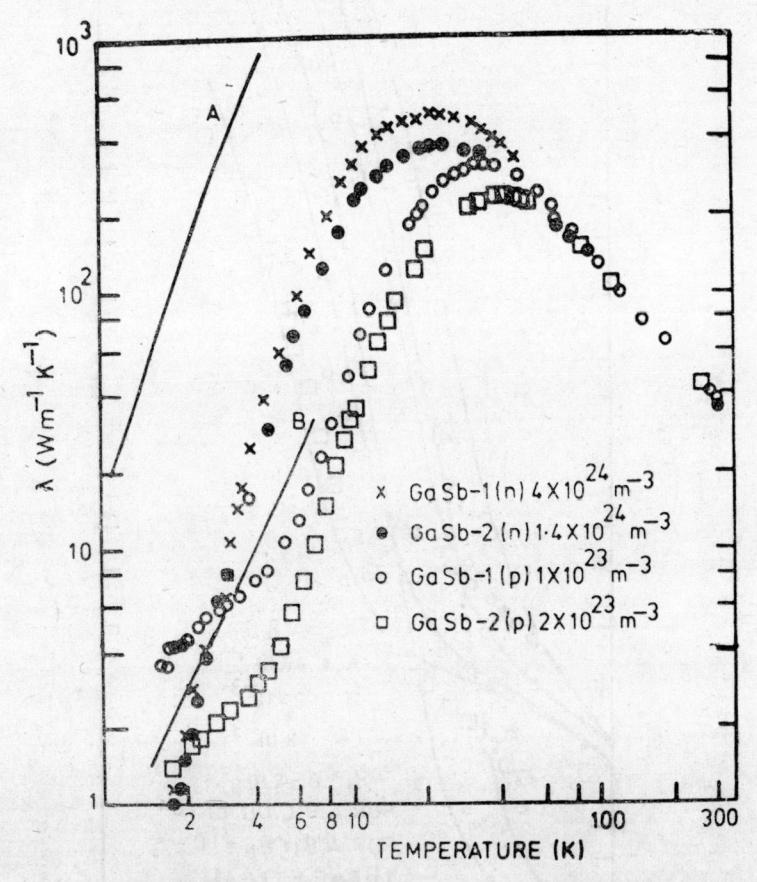

Fig. 8.6 Thermal conductivity of GaSb versus temperature (after Holland, 1964 a). The data marked by the symbol ● (Kopec, 1958) are somewhat lower near the conductivity maximum.

Curve A is the theoretical curve which includes boundary scattering. The reduction in thermal conductivity with doping can be analysed in terms of the additional scattering of phonons by free charge carriers in accordance with Ziman's theory (curve B). This gives the correct magnitude of λ but does not explain the change in the slope of the λ–T curve at around 5 K. The inclusion of phonon scattering by electron (or hole) bound to the donor (or acceptor) impurities (Keyes, 1961, and Griffin and Carruthers, 1963) gives the set of curves shown in Fig. 8.7. These calculations indicate a change in slope around 5 K.

Measurements on heavily doped *p*-type samples were reported by Sota

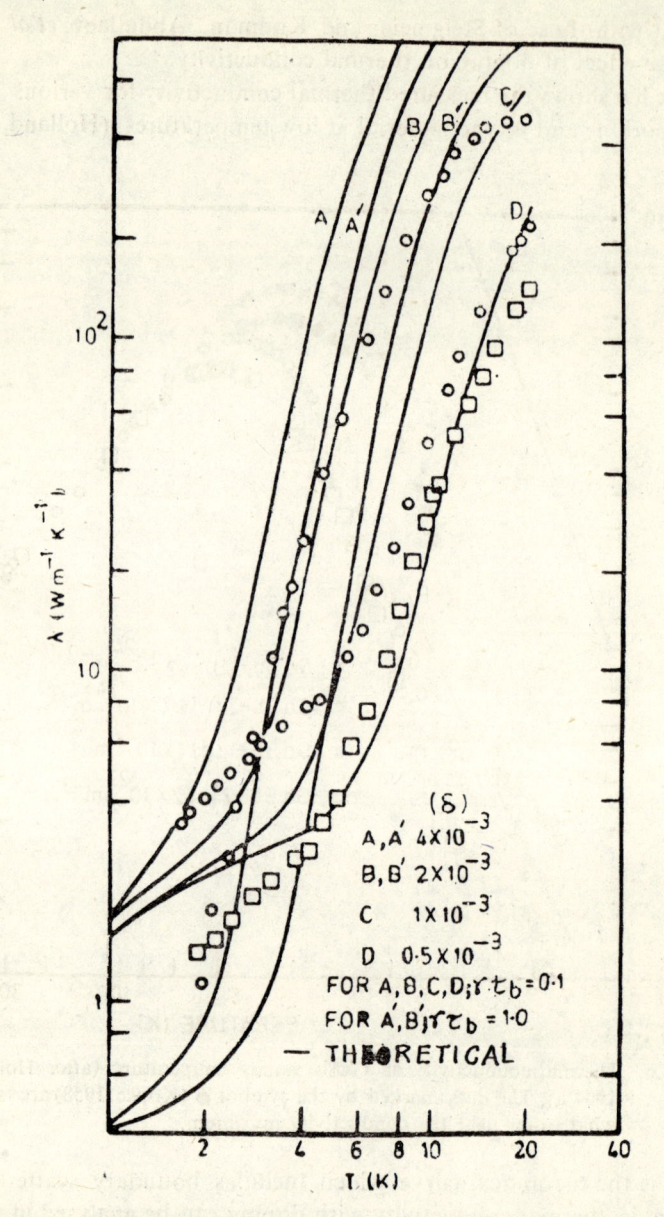

Fig. 8.7 Low-temperature thermal conductivity of GaSb (after Holland, 1964 a). Theoretical curves obtained using $\tau_{pe}^{-1} = \gamma X^4 T^4 (1 + \delta X^2 T^2)^{-8}$ and $\omega \ll \Delta/\hbar$.

et al. (1984). They presented an explanation based on their own theoretical formulation (Sota and Suzuki, 1983) which requires the inclusion of the concentration dependence of shear terms arising from the multi-band structure of the valence bands.

8.2.5 Other III–V compounds

In other III–V compounds, the temperature variation of thermal conductivity can be described on lines similar to those of antimonides and arsenides of indium and gallium described earlier. High-temperature thermal conductivities of III–V compounds (including those of InP, GaP and AlSb) have been plotted in Fig. 8.8, and show an appreciable deviation from the T^{-1} law. This was initially interpreted in terms of the four-phonon proces-

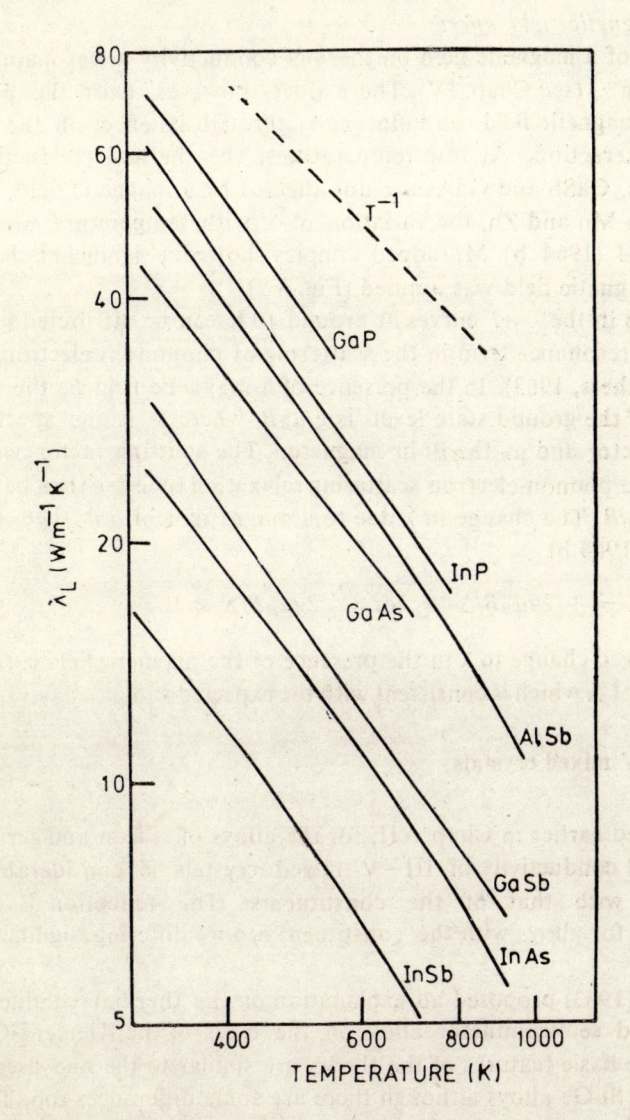

Fig. 8.8 High temperature thermal conductivity of III–V compounds (after Steigmeier, 1969).

ses (Stuckes, 1957, Busch et al., 1959, Glassbrenner and Slack, 1964, Steigmeier and Kudman, 1966) whereas Ranninger (1965) ascribed this deviation to the temperature dependence of the phonon frequency. However, as discussed earlier in Chap. VI, the four-phonon processes are too weak to account for this effect and a satisfactory explanation can be provided by considering the effect of thermal expansion and the additional scattering caused by optical phonons (Slack, 1979).

8.2.6 *Magnetic field effect*

The effect of a magnetic field on thermal conductivity arises mainly due to its effect on λ_e (see Chap. IV). There does, however, exist the possibility where a magnetic field can influence λ_L through its effect on the electron-phonon interaction. At low temperatures, the thermal conductivities of InSb, InAs, GaSb and GaAs are not affected by a magnetic field. In GaAs doped with Mn and Zn, the variation of λ with temperature was studied by Holland (1964 b). Mn-doped samples showed a significant change in λ when a magnetic field was applied (Fig. 8.9).

The dip in the $\lambda - T$ curves at around 10 K can be attributed to the presence of a resonance term in the scattering of phonons by electrons (Griffin and Carruthers, 1963). In the presence of a magnetic field B, the effect of splitting of the ground state levels is $g\mu_B B$ where g is the spectroscopic splitting factor and μ_B the Bohr magneton. The splitting factor which appears in the phonon-electron scattering relaxation time can then be replaced by $\Delta + g\mu_B B$. The change in λ due to a magnetic field B is then given by (Holland, 1964 b)

$$\Delta\lambda/\lambda = \pm\, 2g\mu_B B/\Delta \quad \text{for} \quad 2g\mu_B B/\Delta \ll 1$$

The measured change in λ in the presence of the magnetic field corresponds to a value of Δ which is consistent with the expected value.

8.3 III—V mixed crystals

As described earlier in Chap. VII, for the alloys of silicon and germanium, the thermal conductivity of III—V mixed crystals is considerably lower compared with that of the constituents. The reduction is expected to be large for alloys with the constituent atoms differing significantly in mass.

Abeles (1963) proposed an explanation of the thermal conductivity of a disordered semiconductor alloy on the basis of the Klemens-Callaway theory. The basic features of the theory are similar to the one used in the analysis of Si-Ge alloys although there are some differences too. The large thermal resistance of Si-Ge alloys is predominantly due to the mass-difference scattering, whereas in the ternary alloys of III—V compounds strain-field

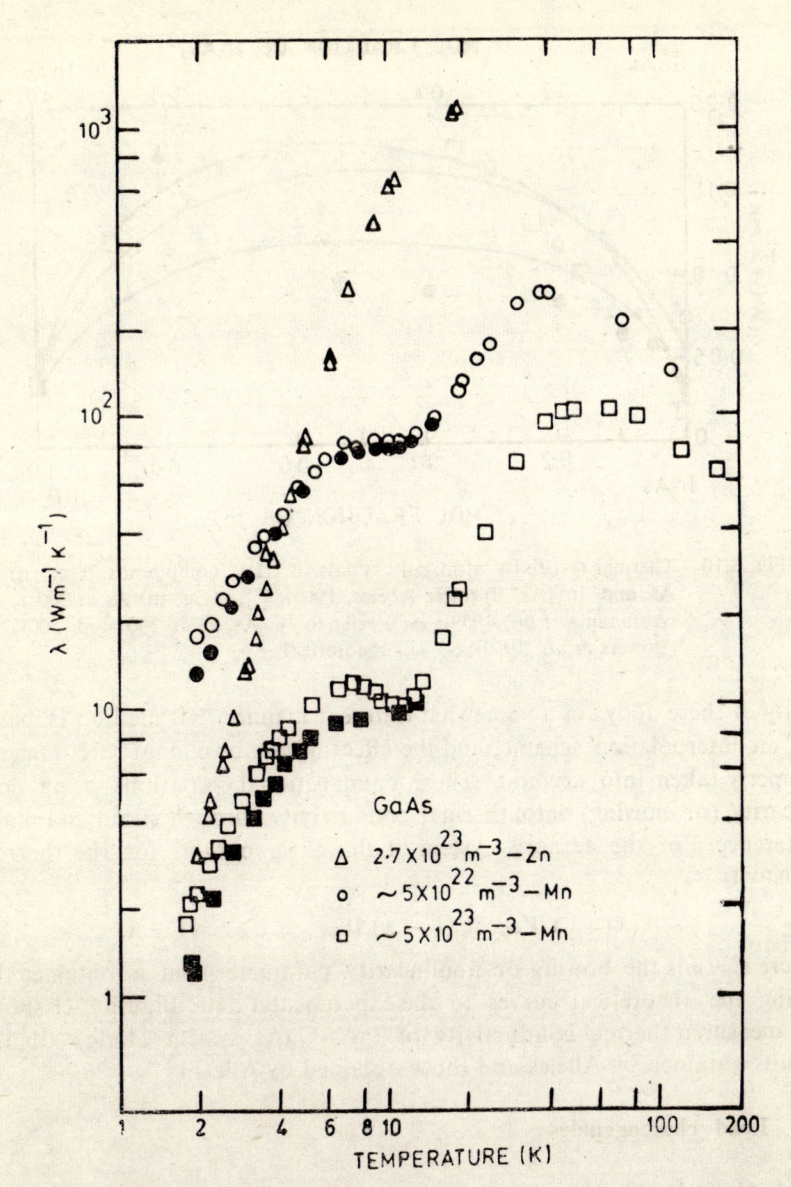

Fig. 8.9 Low temperature thermal conductivity of GaAs-effect of magnetic field (after Holland 1964 b). The closed symbols represent measurements in a magnetic field of 1.27 T.

scattering also plays a significant role. Figure 8.10 gives the thermal resistivity of (Ga, In)As and In(As, P) alloys as a function of composition. To obtain the theoretical curves Abeles used a value of 2.5 for the ratio of three-phonon relaxation times for the Umklapp and Normal processes and several adjustable parameters to obtain the best fit with the observed results.

Adachi (1983) presented a theoretical calculation of the thermal resis-

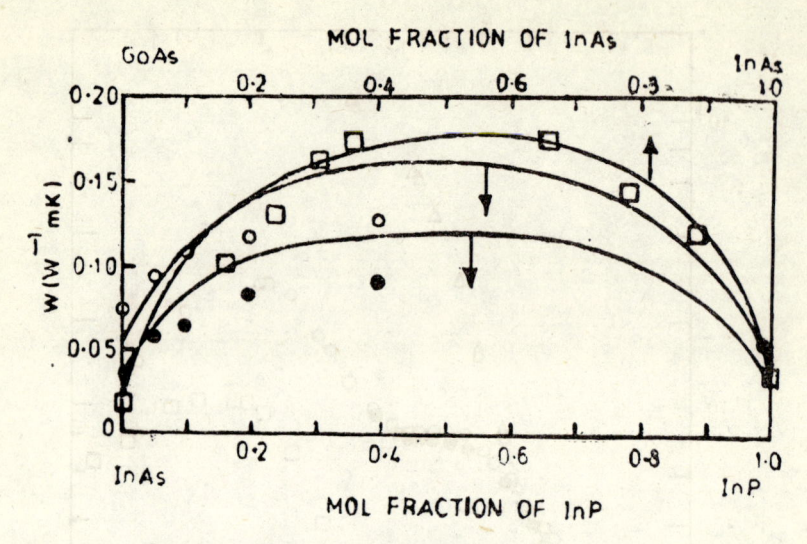

Fig. 8.10 Thermal resistivity of mixed crystals of III-V compounds (Ga, In) As and In (As, P) (after Abeles, 1963). □, (Ga, In) As at 300 K (Abrahams *et al.*, 1959); ●, 0 refer to In (As, P) at 300 and 500K (Bowers *et al.*, 1959); ――― theoretical.

tivity of these alloys in a somewhat different manner. His method is based on an interpolation scheme, and the effect of compositional variations are properly taken into account. These compositional variations bring non-linearity (or bowing) into thermal conductivity through strain and mass-difference. For the ternary system $A_xB_{1-x}C$ he obtains for the thermal resistivity

$$W(x) = xW_{AC} + (1-x)W_{BC} + C_{A-B}x(1-x)$$

where C_{A-B} is the bowing or nonlinearity parameter and is obtained by fitting the theoretical curves to the experimental data. Figure 8.11 shows the measured thermal conductivity of InAs-GaAs system along with the results obtained by Abeles and those obtained by Adachi.

8.4 Lead chalcogenides

8.4.1 Introduction

Studies of the thermal conductivity of lead chalcogenides have been carried out in considerable detail, and the results of these investigations find useful applications in thermoelectric devices. Besides this, they find application in infrared technology; photoresistors made of these materials are used in the measurement and detection of infrared radiation. Photodiodes and lasers based on lead chalcogenides have found useful applications, and in many situations the thermal conductivity of the material is an important parameter.

The electronic thermal conductivity of lead chalcogenides has been studied

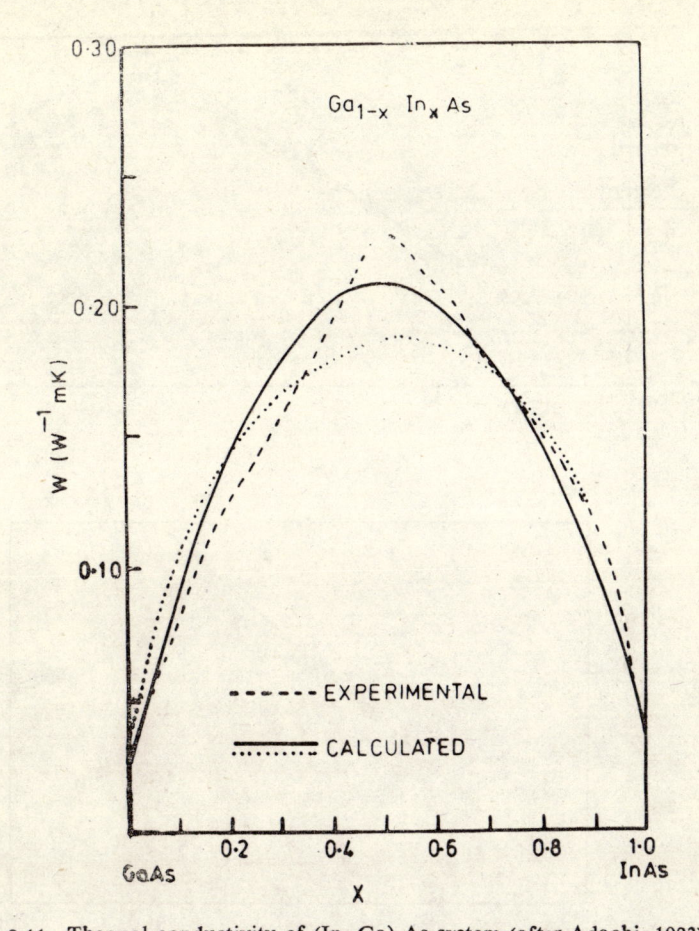

Fig. 8.11 Thermal conductivity of (In, Ga) As system (after Adachi, 1983).

by Smirnov and Ravich (1967) and Ravich *et al* (1971). Theoretical analysis of the electronic contribution to the observed conductivity requires the inclusion of the band non-parabolicity effects on various transport coefficients (see Chap. IV).

8.4.2 *Temperature variation of thermal conductivity*

Figure 8.12 shows the lattice thermal resistivity W_l as a function of temperature for *n*- and *p*-type samples of PbTe and PbSe. The lattice contribution is obtained by subtracting the electronic (polar) contribution λ_e and the photon contribution λ_{photon} from the measured thermal conductivity

$$\lambda_L = \lambda - \lambda_e - \lambda_{photon}$$

Above the Debye temperature, W_l versus T curves can be represented by straight lines. At still higher temperatures there is considerable departure from the linear dependence and this has been attributed to the presence of bipolar contribution (Ravich *et al*, 1970).

162   THERMAL CONDUCTION IN SEMICONDUCTORS

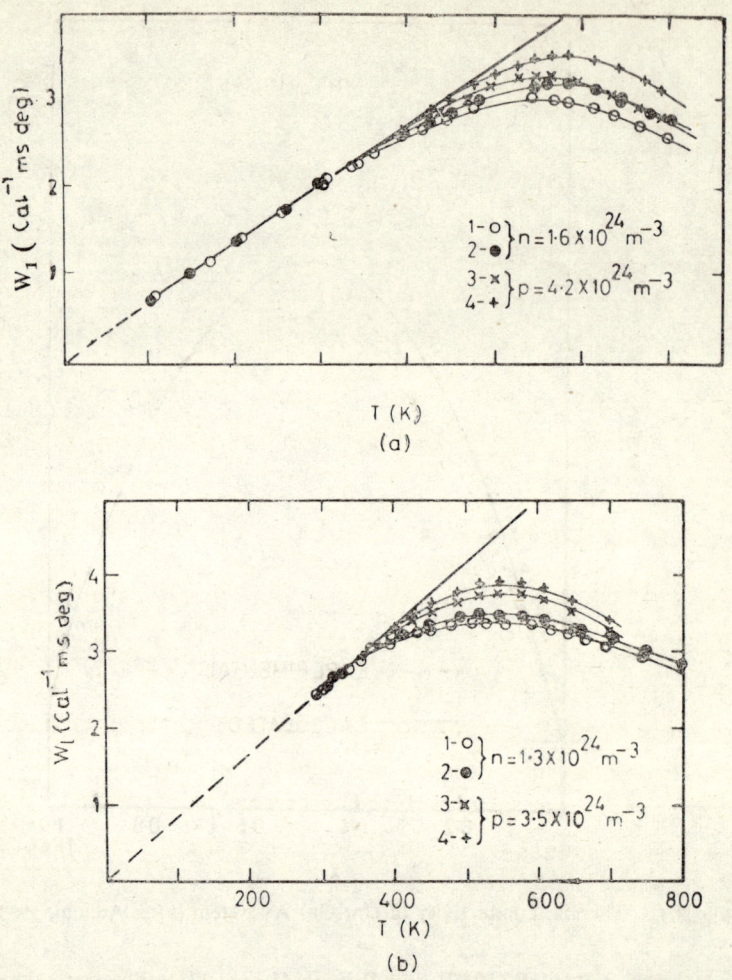

Fig. 8.12 (a) Lattice thermal resistivity of PbTe: 1, 3 (experimental data) $\lambda_{meas} - \lambda_e$; 2, 4 – $\lambda_{meas} - \lambda_e - \lambda_{photon}$ (after Ravich et al., 1970). Dashed curve-extrapolation of high temperature data from Devyatkova and Smirnov (1961).

Fig. 8.12 (b) Lattice thermal resistivity of PbSe: 1, 3-based on experimental data $\lambda_{meas} - \lambda_e$; 2, 4-based on $\lambda_{meas} - \lambda_e - \lambda_{photon}$ (Devyatkova and Saakyan, 1967, Ravich et al., 1970); dashed curve extrapolated from the data of Devyatkova and Smirnov (1961).

In the undoped crystals of PbTe, the photon thermal conductivity is appreciable at sufficiently high temperatures. In doped samples, this contribution is expected to be small due to a high absorption coefficient. In polycrystalline samples λ_{photon} is smaller as photons are scattered at the grain-boundaries (Ravich et al., 1970). Devyatkova and Saakyan (1967) estimated the photon thermal conductivity in PbTe samples with carrier concentrations around to^{26} m^{-3} and found it to be about 4–5 percent of the total thermal conductivity at 560 K.

The thermal conductivity of PbTe increases with the carrier concentration (Devyatkova and Smirnov, 1961) and this increase is due to the electronic contribution (λ_e) as the lattice contribution (λ_L), which varies approximately as T^{-1}, is reduced slightly with doping. Devyatkova and Smirnov (1962) ascribed this reduction in λ_L to the scattering of phonons by impurity atoms. Steigmeier (1969) pointed out that the reduction in λ_L is essentially caused by the scattering of phonons by free carriers (see Sec. 7.3). Bhandari and Rowe (1984) presented an analysis of the thermal conductivity data of PbTe by considering both the acoustic phonon scattering and polar optical mode scattering of the carriers including non-parabolicity of the bands and a lattice contribution which includes the phonon scattering by free carriers.

8.4.3 Mixed crystals

Mixed crystals of lead chalcogenides have found applications in thermoelectric devices. The performance of thermoelectric devices can be improved if thermal conductivity could be reduced without being accompanied by a degradation in electrical properties. Ioffe and co-workers have shown that this can be achieved by preparing mixed crystals Pb $X_a Y_{1-a}$, where X and Y stand for S, Se or Te. The lattice thermal resistivity of a PbTe — PbSe alloy (50–50 composition) increases by a factor of two over the resistivity of PbTe or PbSe (Ioffe, 1956, and Ioffe and Ioffe, 1956). Bhandari and Rowe (1983) have studied the effect of grain-boundary scattering on the thermal conductivity of fine-grained lead telluride. The effect of grain-boundary scattering is found to be considerably enhanced in highly disordered alloys.

Measurement of the thermal conductivity of PbTe and its solid solutions with SnTe, MnTe and PbSe have been carried out (Devyatkova and Smirnov, 1961, and Ravich et al., 1967, 1970, and 1971) in considerable detail. The electronic contribution to λ which is influenced by band non-parabolicity must be subtracted from the measured thermal conductivity to obtain the lattice contribution. Ravich et al (1967) analysed the results of measurements of λ of PbTe and PbTe-PbSe solid solutions in the temperature range 80–300 K. They pointed out that the Lorenz number L (see Chap. IV) divided by L_0 (the value of L in the degenerate limit) is 1 at 4 K for samples with carrier concentration 10^{24} m^{-3}, but decreases with increasing temperature. Around 77 K it acquires a minimum value of $\simeq 0.6$ and then increases to unity at 300 K. For higher carrier concentrations ($n \simeq 10^{25}$ m^{-3}), $L/L_0 = 1$ in the entire range 80–300 K. For PbTe-PbSe solid solutions ($n = 10^{24}$ m^{-3}) the solutions with majority of PbTe behave as PbTe. If PbSe content is large the ratio is unity in the whole range of temperature considered. Figure 8.13 gives the variation of L/L_0 with temperature for PbTe and its solid solutions (Ravich et al, 1967).

164 THERMAL CONDUCTION

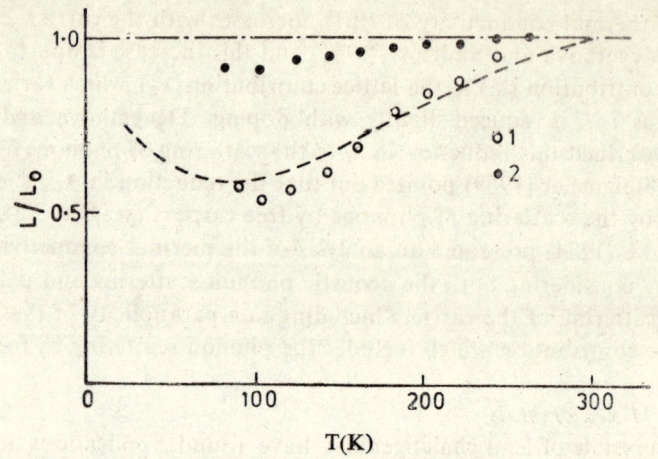

Fig. 8.13 Temperature dependence of L/L_0 for PbTe-PbSe solid solutions. (1) 5% PbSe and (2) 40% PbSe. Lower dashed curve refers to PbTe with $n = 10^{24}$ m^{-3}.
(after Ravich et al., 1967).

8.5 II-VI compounds

Thermal conductivity measurements were made by Slack and Galginaitis

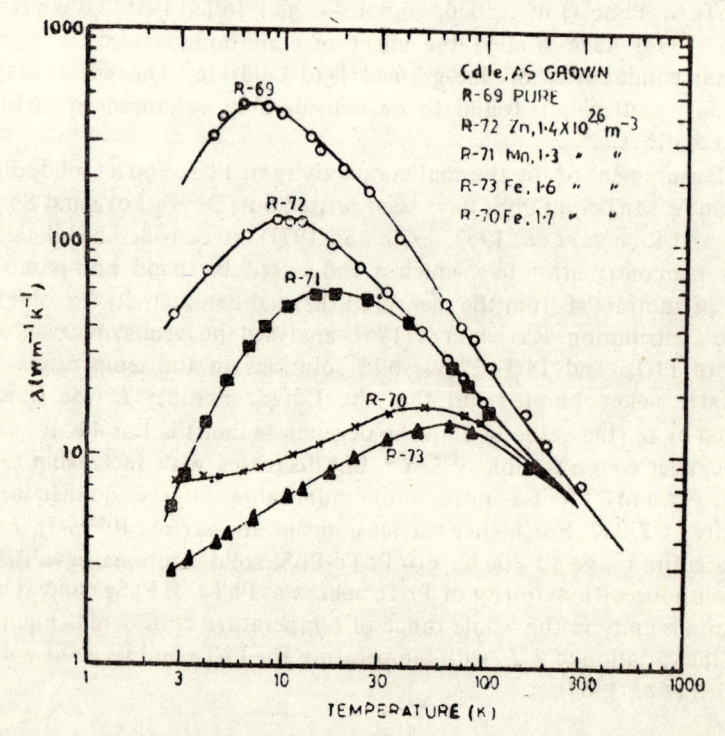

Fig. 8.14 Thermal conductivity versus temperature for pure and doped CdTe containing about 1 mol % impurity. The crystals were measured in the as grown state before any heat treatment (after Slack and Galginaitis, 1964).

(1964) on CdTe crystals doped with Mn, Fe or Zn. Heat transport in pure CdTe is essentially due to phonons and the phonon mean-free-path is limited mainly by other phonons, isotopes and crystal boundaries. Theoretical calculations based on the Callaway model are found to show good agreement with experimental results in pure samples below 80 K (Holland, 1964 a). An additional phonon scattering is present in all doped crystals but a resonant type of scattering substantially reduces λ in Fe-doped crystals. This is ascribed to a resonant scattering of phonons from the low lying states of Fe^{++} ions. Samples doped with Zn and Mn do not show this type of behaviour (Fig. 8 14).

Slack (1972) reported measurements of λ on ZnO, ZnS, ZnSe, ZnTe and CdTe in the temperature range 3–300 K. The undoped samples of all these compounds exhibited similar behaviour. In these and certain other materials with adamantine structure, the room temperature thermal conductivity can be analysed in terms of a phonon contribution which is limited mainly by phonon-phonon interaction. The thermal conductivity is then written as

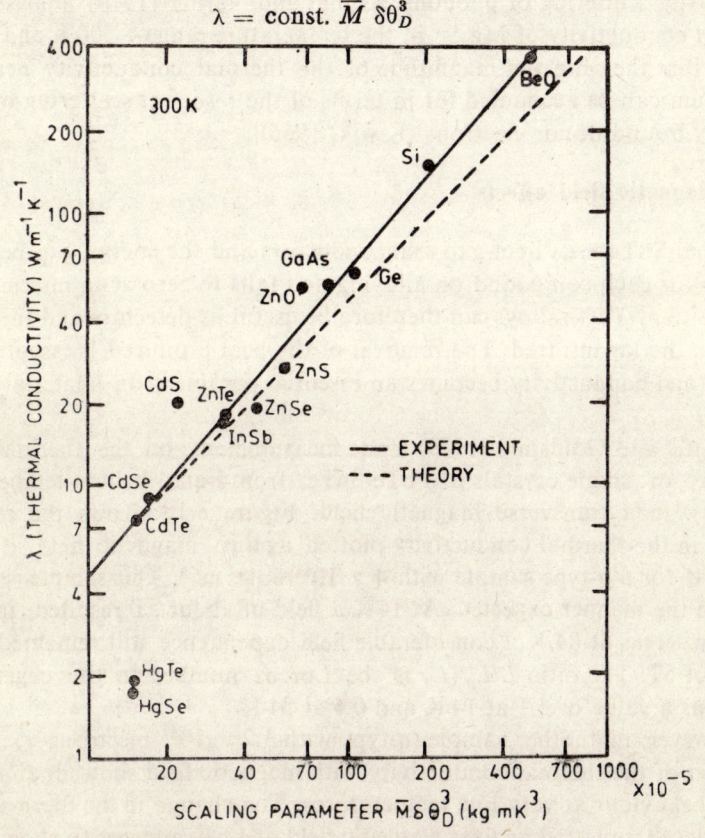

$$\lambda = \text{const.}\ \overline{M}\ \delta\theta_D^3$$

Fig. 8.15 Room temperature thermal conductivity of II–VI compounds along with some other semiconductors plotted against the scaling factor $\overline{M}\delta\theta_D^3$. Dashed line shows the results of theoretical calculations for $\gamma = 2$ (after Slack, 1972).

where θ_D is the Debye temperature extrapolated to absolute zero, δ^3 is the average atomic volume and $\overline{M}$ is the average atomic mass. Figure 8.15 shows the measured thermal conductivity at 300 K plotted against the scaling parameter $\overline{M}\,\delta\theta_D^3$. With the exception of HgSe and HgTe most of the compounds fit fairly well on a straight line. The calculated values for $\gamma = 2$ are also shown.

8.6 II–IV compounds

The thermal conductivities of II–IV compounds (Mg_2Sn, Mg_2Si and Mg_2Ge) have been measured by Martin and Danielson (1968) and Martin (1972) from liquid helium temperature to room temperature. Martin *et al* (1968 and 1969) reported measurements of λ in Mg_2Sn and Mg_2Ge up to 700 K. The observed data were analysed in terms of Holland's model in the temperature range 4–700 K (Martin, 1972). Samples with carrier concentrations above $4 \times 10^{22}\,m^{-3}$ were reported to show neutral donor scattering and also free-carrier scattering of phonons. Kumar and Verma (1970) analysed the thermal conductivity of Mg_2Sn in the temperature range 4–300 K and pointed out that the observed magnitude of the thermal conductivity near the maximum can be accounted for in terms of the resonant scattering of phonons by bound donor electrons (Kwok, 1966).

8.7 Magnetic field effects

PbTe and SnTe are direct-gap semiconductors and the energy gap becomes smaller for each compound on alloying and falls to zero at an intermediate composition. These alloys can therefore be useful as detectors and injection lasers in the far infrared. The removal of the heat produced is essential and the thermal conductivity becomes an essential parameter in material selection.

Knittel and Goldsmid (1979) made measurements on the thermal conductivity of single crystals of PbTe-SnTe, from liquid helium temperature upwards, in a transverse magnetic field. Figure 8.16 shows the relative change in the thermal conductivity plotted against magnetic field at 14 K and 84 K for a *p*-type sample with 4×10^{24} holes m^{-3}. This sample behaved much in the manner expected. At 14 K a field of about 5T resulted in saturation whereas at 84 K a considerable field dependence still remained with a field of 5T. The ratio L/L_0 (L_0 is the Lorenz number in the degenerate limit) has a value of 1.1 at 14 K and 0.6 at 84 K.

However, in another sample (*n*-type with 3.2×10^{22} electrons m^{-3}) the variation of the thermal conductivity with magnetic field showed an unexpected behaviour at very low temperatures. The change in the thermal conductivity ($\Delta\lambda$) plotted against magnetic field at 14 K appears to show saturation around 1.5 T (Fig. 8.17 (a)) but then increases at higher fields. At a still lower temperature (5.3 K) the $\Delta\lambda$ versus B curve shows oscillatory behaviour (Fig. 8.17 (b)).

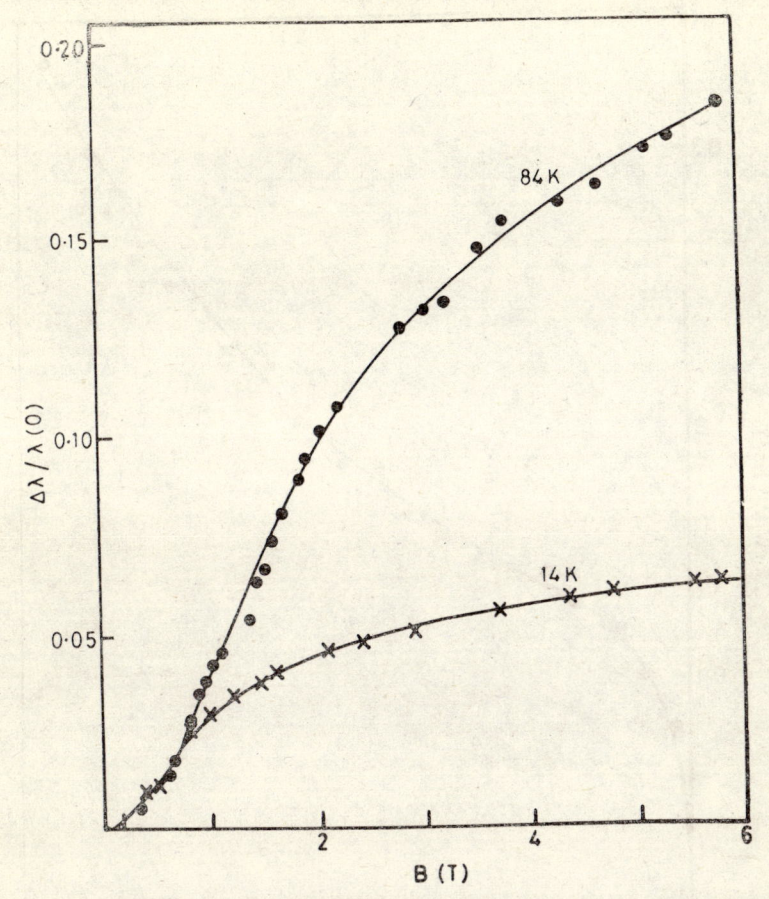

Fig. 8.16 Relative change in the thermal conductivity of PbTe-SnTe solid solution plotted against the magnetic field at 14 K and 84 K (after Knittel and Goldsmid, 1979).

The change in the thermal conductivity shown by the second sample corresponds to a value of the Lorenz number L which is greater than L_0. At these low temperatures there is no possibility of bipolar thermal conduction. Knittel and Goldsmid (1979) suggested an explanation in terms of a phonon-drag contribution to the magnetothermal resistance. The oscillatory behaviour can be explained on the basis of the periodic dependence of the electron density-of-states on the magnetic field (Bresler et al, 1966).

8.8 Bismuth telluride and other V–VI compounds

Bismuth telluride has a rhombohedral crystal structure and the atoms are arranged in layers with like atoms arranged in the rhombohedral [111] direction parallel to the c-axis. The crystals can be readily cleaved along planes perpendicular to the c-axis and there exists a strong anisotropy in the

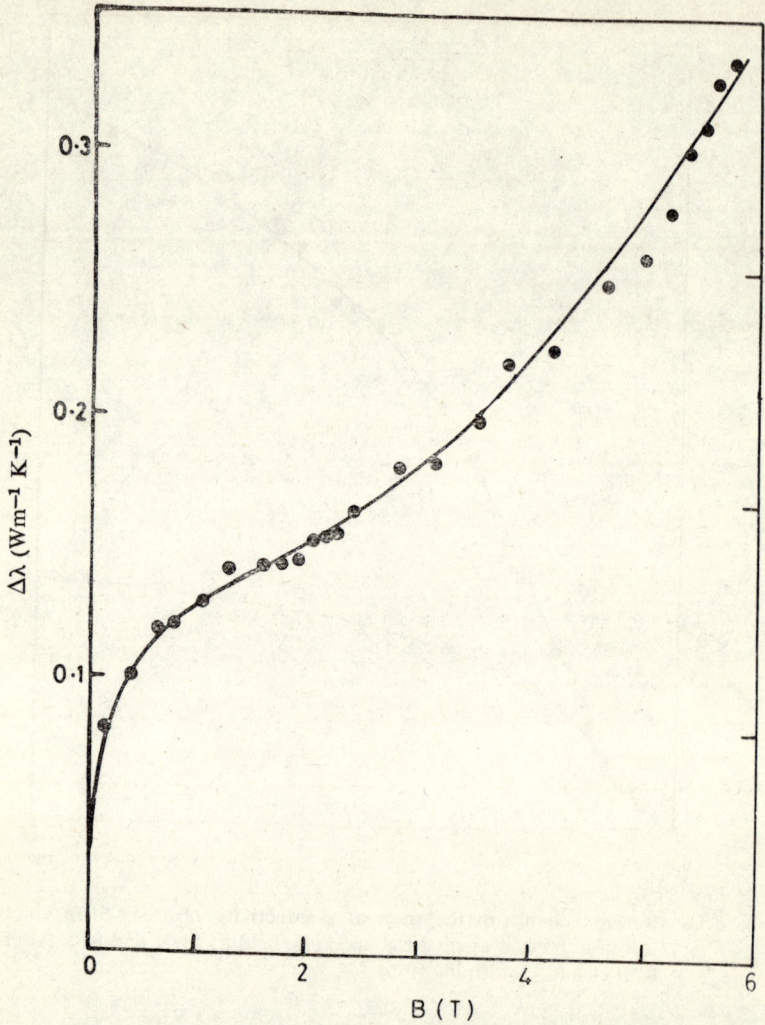

Fig. 8.17 (a) $\Delta\lambda$ versus B curve for an n-type sample of PbTe-SnTe alloy at 14 K (after Knittel and Goldsmid, 1979)

various transport properties including thermal conduction. At 300 K the thermal conductivity of single crystal Bi_2Te_3 in a direction parallel to the cleavage planes (λ_{11}) is about 1.5 $Wm^{-1}K^{-1}$ and that along a direction perpendicular to the cleavage planes ($\lambda_\perp$) is 0.7 $Wm^{-1}K^{-1}$ (Goldsmid, 1965). The study of the thermal conductivity of anisotropic crystals requires good single crystals. Most of the measurements have been made on polycrystalline specimens.

Goldsmid (1956) reported measurement of the thermal conductivity of crystals cut parallel and perpendicular to the c-axis between 130 K and 300 K. Measurements by Walker (1960) in the temperature range 6–200 K compare well with Goldsmid's (1958) values at 150 K. Between 50 and 150 K

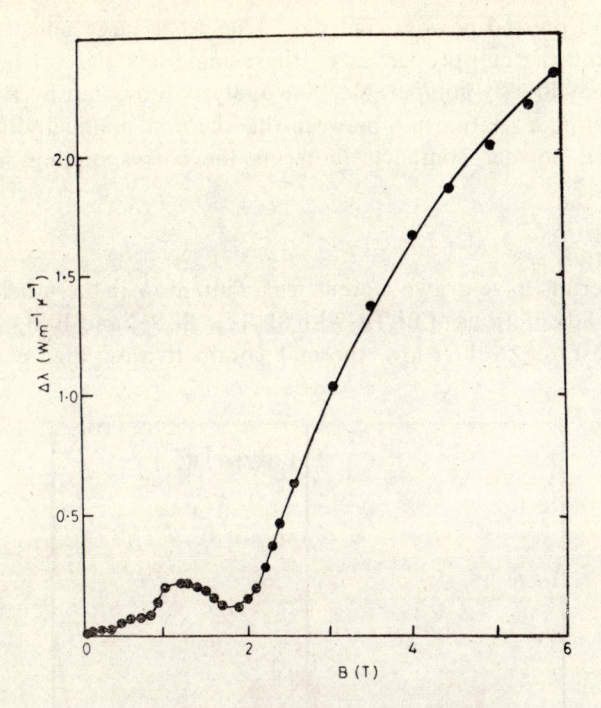

Fig. 8.17 (b) $\Delta\lambda$ versus B for PbTe-SnTe alloy at 5.3 K (after Knittel and Goldsmid, 1979).

the thermal conductivity was found to vary as T^{-1}. At higher temperatures the dependence was weaker than T^{-1} and this can be ascribed to the electronic component. The correlation between the thermal and electrical conductivities on doped samples of Bi_2Te_3 is shown in Fig. 4.4 (see chapter IV) and a significant electronic contribution to the thermal conductivity is apparent. The dependence of λ on doping (for halogen-doped samples) was interpreted by Goldsmid in terms of the scattering of phonons by impurities. However, phonon-electron scattering might also be important and can be calculated in a way similar to that discussed for other semiconductors (Steigmeier and Abeles, 1964). The electronic thermal conductivity may be strongly influenced by the nonparabolic nature of the energy bands (see Chap. IV, and also Bhandari and Rowe, 1984).

Satterthwaite and Ure (1957) also measured thermal conductivity of bismuth telluride and found a T^{-1} variation in the range 80–170 K. Their values were about 20% higher than those reported by Walker.

The thermal conductivity of Bi_2Se_3 was measured by MacDonald et al. (1959) and showed a maximum around 10 K. Hashimoto (1958) also measured λ and found that it differed little from that of bismuth telluride in the temperature range 110–250 K.

Measurements on the thermal conductivity of bismuth telluride and other V–VI semiconductors have been made on single crystals, polycrystalline samples and pressed powder compacts. Due to a large anisotropy in the electrical and thermal properties of these materials the various measurements are not directly comparable. The analysis presented by Ryden (1971) aims at finding a relationship between the thermal conductivities of polycrystals and powder compacts in terms the corresponding single-crystal values.

Mixed crystals
These materials have drawn a great deal of attention in thermoelectric applications. Solid solutions of Bi_2Te_3 with Sb_2Te_3, Bi_2Se_3 and Bi_2S_3 are of particular interest due to their low thermal conductivities. Figure 8.18 shows

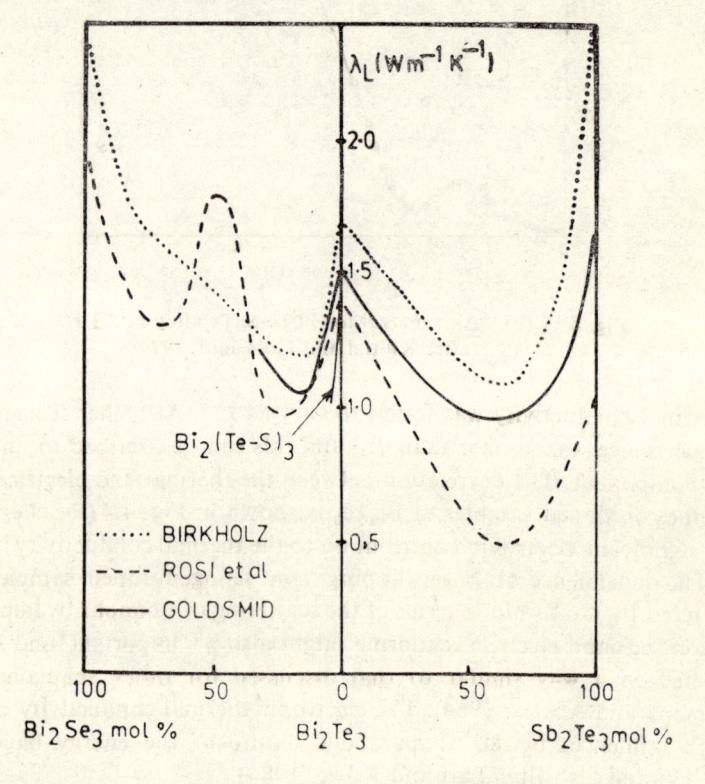

Fig. 8.18 Room temperature thermal conductivity of mixed compounds $Bi_2Te_3-Bi_2Se_3$ and $Bi_2Te_3-Sb_2Te_3$ (after Goldsmid, 1965)

the variation of λ_L for some of the solid solutions along the cleavage planes. The results obtained by various authors (Birkholz, 1958, Rosi *et al*, 1959 and Goldsmid, 1961) disagree over certain ranges of composition in $Bi_2(Te, Se)_3$ system. In the rest of the composition range for this and other system (Bi, Sb)$_2$ Te$_3$ there is qualitative agreement and quantitative differences can be attributed to the error in estimating the electronic contribution.

The variation of the lattice thermal conductivity in the $(Bi, Sb)_2Te_3$ system with composition can be explained reasonably well on the basis of the point-defect scattering arising solely due to the mass-difference between Sb and Bi atoms. For the (Te–S) and (Te–Se) systems λ falls more rapidly with the addition of Bi_2S_3 or Bi_2Se_3 to Bi_2Te_3 and this has been attributed (Goldsmid, 1965) to disturbances in the forces between atoms.

The anisotropy ratio for the lattice thermal conductivity, however, is found to be almost the same for alloys as well as pure Bi_2Te_3. Blitz et al. (1960) have attributed this strong anisotropy to the anisotropy of the phonon velocity.

Order-disorder transformation
Some solid solutions show a transition to an ordered state (Abrikosov and Bankina, 1979). The thermal conductivity of the systems Sb_2Te_3–Bi_2Se_3, Sb_2Te_3–Sb_2Se_3 and Sb_2Te_3–Bi_2Te_3 have been investigated after annealing. For a two-to-one ratio in the first two systems and two-to-one and one-to-two ratios in the third there is a decrease in the thermal conductivity during the first hours of annealing. This may be due to stresses in the crystal lattice induced by the rearrangement of the latter. With further annealing there is an increase in the thermal conductivity which finally acquires the value corresponding to an ordered state. For other compositions such $Bi_2Te_3 \cdot 3Sb_2Te_3$ or $4Sb_2Te_3 \cdot Sb_2Se_3$, no such decrease in the thermal conductivity is observed, which suggests an absence of ordered structure.

8.9 Other semiconductors

Most of the research in semiconductors has been confined to tetrahedral semiconductors. Among other materials, lead chalcogenides have been investigated thoroughly. Apart from these and others discussed so far, a great variety of semiconductors exist (Goodman, 1958, and 1984); several minerals which include pyrites, beegerite, chalcopyrite and intermediate phases of systems, such as PbS–Sb_2S_3, show semiconducting properties. These materials obey normal chemical-valence rules and many of them donot have tetrahedral bonding. In many of these materials two different types of bonds are present—for example in $CuFeS_2$, Fe–S and Cu–S bonds. This corresponds to energy gaps in parallel, the effective gap of the material being determined by the weaker bond (Goodman, 1984).

The search for new materials for a variety of applications (see chapter XI) led to a detailed study of various minerals, some of which have been described above (Goodman, 1984). Thermal conductivity measurements on these materials and the analysis of the data is likely to draw a greater attention of physicists and device engineers in the coming years.

References

Abeles, B. (1963), *Phys. Rev. 131*, 1906.
Abdullaev, G.B., Aliev, G.M. and Safaraliev, G.I. (1966), *Phys. Stat. Solidi 17*, 777.
Abrahams, M.S., Braunstein, R. and Rosi, F.D. (1959), *J. Phys. Chem. Solids 10*, 204.
Abrikosov, N.K. and Bankina, V.F. (1979), *Inorganic Matter 15*, No. 6, 863.
Adachi, S. (1983), *J. Appl. Phys. 54*, 1844.
Amirkhanov, K.I. and Magomedov, V.B. (1966), *Sov. Phys.-Solid St. 8*, 241.
Amirkhanov, D.K. and Bashirov, R.I. (1960), *Sov. Phys.-Solid St. 2*, 1447.
Amith, A., Kudman, I. and Steigmeier, E.F. (1965), *Phys. Rev. 138*, A 1270.
Beers, D.S., Cody, G.D. and Abeles, B. (1962), *Proc. of Int. Conf. on Physics of Semicond.* Exeter, The Inst. of Physics aud Physical Society.
Bettman, M. and Schneider, J.E. (1960), *Thermoelectricity* (ed. P.H. Egli), Wiley, New York.
Bhandari, C.M. and Verma, G.S. (1965), *Phys. Rev. 140*, A 2101.
Bhandari, C.M. and Rowe, D.M. (1983), *J. Phys. D: Appl. Phys. 16*, L 75.
Bhandari, C.M. and Rowe, D.M. (1984), *Ninth European Conf. Thermophys. Properties*, Manchester, Sept. 1984.
Birkholz, U. (1958) *Z Nathurforsch 13a*, 780.
Blitz, J., Clunie, D.M. and Hogarth, C.A. (1960), *Proc. of. Int. Conf. on Phys. of Semicond.*, Prague, 1960, p. 641. Czechoslovak Acad. of Sciences, Prague.
Bowers, R., Ure, R.W., Bauerle, J.E. and Cornish, A.J. (1959), *J. Appl. Phys. 30*, 930.
Bresler, M.S., Parfenev, R.V. and Shalyt, S.S. (1966), *Sov. Phys.-Solid St. 8*, 1414.
Busch, G. and Schneider, W. (1954), *Helv. Phys. Acta 27*, 196.
Busch, G. and Schneider, W. (1954), *Physica 20*, 1084.
Busch, G., Steigmeier, E.F. and Wettstein, E. (1959), *Helv. Phys. Acta 32*, 463.
Busch, G. and Steigmeier, E.F. (1961), *Helv. Phys. Acta 34*, 1
Challis, C.J., Cheeke, J.D. and Williams, D.J. (1964), *Proc. of 9th Int. Conf. on Low Temp. Phys.*, Plenum Press, New York, p. 1145.
Devyatkova, E.D., Moizhes, B.Y. and Smirnov, I.A. (1959), *Sov. Phys.-Solid St. 1*, 555.
Devyatkova, E.D. and Saakyan, V.A. (1967), *Izv. Akad. Arm. SSR, Ser. Fiz. 2*, 14.
Devyatkova, E.D. and Smirnov, I.A. (1961), *Sov. Phys.-Solid St. 3*, 1666.
Devyatkova, E.D. and Smirnov, I A. (1962), *Sov. Phys.-Solid St. 4*, 1836.
Gaur, N.K.S., Bhandari, C.M. and Verma, G.S. (1966), *Physica 32*, 1048.
Glassbrenner, C.J. and Slack, G.A. (1964), *Phys. Rev. 134*, A1058.
Goldsmid, H.J. (1954), *Proc. Phys. Soc. B 67*, 360.
Goldsmid, H.J. (1956), *Proc. Phys. B 69*, 203.
Goldsmid, H.J. (1958), *Proc. Phys. Soc. 72*, 17.
Goldsmid, H.J. (1961), *J. Appl. Phys. 32*, 2198.
Goldsmid, H.J. (1965) *Materials Used in Semiconductor Devices*, (ed. C.A. Hogarth), Interscience Publ.
Goodman, C.H.L. (1958), *J. Phys. Chem. Solids 6*, 305.
Goodman, C.H.L. (1984), *Int. Conf. on Solid St. Devices and Materials*, Kobe, Japan, August 1984.
Griffin, A. and Carruthers, P. (1963), *Phys. Rev. 131*, 1976.
Guthrie, G.L. and Whitefield, R.J. (1966), *Bull. Amer. Phys. Soc. 11*, 900.

Hashimoto, K. (1958), *Mem. Eac. Sci.*, Kyushu Univ. B2, 187.

Holland, M.G. (1964 a), *Phy. Rev. 134*, A471.

Holland, M.G. (1964 b), *Physics of Semicond.*, *Proc. Int. Conf. on Semicond.*, Paris. Dunod, Academic Press.

Holland, M.G. (1966), *Physics of III–V Compounds* (ed. R.K. Willardson and A.C. Beer), Vol. 2, Acad. Press.

Ioffe, A.F. (1956), *Can. J Phys. 34*, 1342.

Ioffe, A.V. and Ioffe, A.F. (1956), *Izvest. Akad. Nauk. SSSR 20*, 65.

Kanai, Y. and Nii, R. (1959), *J. Phys. Chem. Solids 8*, 338.

Keyes, R.W. (1961), *Phys. Rev. 122*, 1171.

Knittel, T. and Goldsmid, H.J. (1979), *J. Phys. C: Solid. St. Phys. 12*, 1891.

Kopec, Z. (1958), *Acta Phys. Polon. 17*, 265.

Kumar, A. and Verma, G.S. (1970), *Phys. Rev B1*, 488.

Kwok, P.C. (1966), *Phys. Rev. 149*, 666.

MacDonald, D.K.C., Mooser, E., Pearson, W.B., Templeton, I.N. and Woods, S.B. (1959), *Phil. Mag. 4*, 433.

Martin, J.J. (1972), *J. Phys. Chem. Solids 33*, 1139.

Martin, J.J. and Danielson, G.C. (1968), *Phys. Rev. 166*, 879

Martin, J J., Shanks, H.R. and Danielson, G.C. (1968), **Thermal Conductivity**, *Proc. of 7th Conf.* (ed. D.R. Flynn and B.A. Peavy, Jr.), p. 381.

Martin, J.J., Shanks, H.R. and Danielson, G.C. (1969), **Thermal Cond.**, *Proc. of 8th Conf.* (ed. C.Y. Ho and R.C. Taylor), Plenum Press, p. 795.

Mielczarek, E.V. and Frederikse, H.P.R. (1959), *Phys. Rev. 115*, 888.

Pyle, E.C. (1961), *Phil. Mag. 6*, 609.

Ranninger, J. (1965), *Phys. Rev 140*, A2031.

Ravich, Yu.I., Smirnov, I.A. and Tikhonov, V.V. (1967), *Sov. Phys.-Semicond. 1*, 163.

Ravich., Yu.I., Efimova, B.A. and Smirnov, I.A. (1970), *Semiconducting Lead Chalcogenides*, (ed. L.S. Stil'bans) Plenum Press, New York.

Ravich, Yu.I., Efimova, B.A. and Tamarchenko, V.I. (1971), *Phys. Stat. Solidi (b) 43*, 11.

Rosi, F.D., Abeles, B. and Jensen, R.V. (1959), *J. Phys. Chem. Solids 10*, 191.

Ryden, D.J. (1971), *J. Phys. C: Solid St. Phys. 4*, 1193.

Satterthwaite, C.D. and Ure, R.W. (1957), *Phys. Rev. 108*, 1164

Slack, G.A. (1972), *Phys. Rev. B6*, 3791.

Slack, G.A. (1979), *Solid State Physics* (ed. Ehrenreich *et al*,) Acad. Press, New York, Vol. 34, p. 1.

Slack, G.A. and Galginaitis, S. (1964), *Phys. Rev. 133*, A253.

Smirnov, I.A. and Ravich, Yu.I. (1967), *Sov. Phys.–Semiconductors 1*, 739.

Sota, T. and Suzuki, K. (1983), *J. Phys. C: Solid St. Phys. 16*, 4347.

Sota, T. and Suzuki, K. (1984), *J. Phys. C: Solid St. Phys. 17*, 2661.

Sota, T., Suzuki, K. and Fortier, D (1984), *J. Phys. C: Solid St. Phys. 17*, 5935.

Srivastava, G.P. (1981), *J. Phys. Colloq. 42*. C6–149.

Steigmeier, E.F. and Abeles, B. (1964), *Phys. Rev. 136*, A1149.

Steigmeier, E.F. and Kudman, I. (1963), *Phys. Rev. 132*, 508.

Steigmeier, E.F. and Kudman, I. (1966), *Phys. Rev. 141*, 767.

Steigmeier, E.F. (1969), *Thermal Conductivity* (ed. R.P. Tye), Acad. Press, London.

Stuckes, A.D. (1957), *Phys. Rev. 107*, 427.

Timberlake, A.B., Davis, P.W. and Shilliday, T.S. (1962), *Adv. Ener. Conv. 2*, **45**.
Timberlake, A.B., Davis, P.W. and Shilliday, D.S. (1962), *J. Appl. Phys. 33*, 765.
Wagini, H. (1964), *Z. Naturf. 19a*, 1541.
Walker, B.A. (1960), *Proc. Phys. Soc. London 76*, 113.
Weiss, H. (1958), Semiconductors and Phosphors, *Vieweg*, Braunschweig, p. 497.
Zhuze, V.B. (1954), *Dokl. Akad. Nauk. SSSR*, *98*, 711.
Ziman, J.M. (1956), *Phil. Mag. 1*, 191.
Ziman, J.M. (1957), *Phil. Mag. 2*, 292.

Chapter IX

Amorphous and Liquid Semiconductors

9.1 Amorphous materials

9.1.1 *Introduction*
The semiconductors discussed in previous chapters were crystalline materials and the theory of lattice thermal conductivity developed in Chap. VI had as its basis the regular periodic arrangement of atoms—a "perfect crystal". Any deviation from this perfect periodicity was treated as an imperfection which led to the scattering of the carriers of heat (mainly phonons) and gave rise to a thermal resistance. However, whatever the degree of disorder, the model based on a regular lattice structure remained applicable, and this facilitated an understanding of the various physical properties of the material. Nevertheless, there are certain categories of semiconductors—amorphous materials, where the long-range order is absent and the model based upon a crystalline structure is not a completely valid one.

9.1.2 *Basic features*
The mechanism of heat transport in crystalline material is inadequate to explain the phenomenon of heat transport in amorphous material. The same it true of various electronic properties. In amorphous materials the Bloch type of electronic states are modified to a form of extended states—states for which the electronic wavefunction extends through a region of macroscopic dimensions (Anderson, 1958, Mott, 1967, and Cohen *et al*, 1969). A comprehensive treatment of the subject is given by Gubanov (1965), Mott and Davis (1971), Cohen and Lucovsky (1972) and Tauc (1974).

The vibrational properties of non-crystalline solids have been reviewed by Hori (1968), Bell (1972), Dean (1972) and Bottger (1974). The phonon spectrum in amorphous materials differs from that of the corresponding crystalline material with the high-frequency part of the phonon density-of-states curve having a washed-out appearance as compared to the crystalline density-of-states curves. This behaviour appears to point to the important role played by short-range order. The experimental data indicate an enhanced density-of-states at low frequencies in amorphous materials compared to the crystalline state. These low-frequency modes may be responsible for some of the observed differences between the thermal properties of crystalline and amorphous solids.

Among the materials characterized by an absence of long-range order are glasses, polymers and vitreous materials. Liquid semiconductors can also be placed in this class although a number of their properties differ significantly from those of amorphous materials.

Although there are difficulties in determining the normal modes of vibration in an amorphous material, in principle such modes may be assumed to exist. Some important features of the vibrational properties of amorphous materials are listed below (Klemens, 1965).

(a) At high temperatures the specific heat of these materials is not widely different from that of the crystalline solids. On this basis one may ascribe a similar frequency spectrum for the normal modes except at high frequencies. High-frequency normal modes differ considerably in character from the crystalline modes.

(b) In the long-wavelength limit, the atomic structure is unimportant, and, therefore, low-frequency normal modes must be similar to the waves in an elastic continuum.

(c) Glasses differ from liquids in that liquids cannot propagate low-frequency transverse waves while amorphous solids can.

9.1.3 *Classification of amorphous solids*

Crystalline solids can be classified by the type of chemical bonding. The five conventional classes based on chemical bonding are: ionic, covalent, metallic, Van der Waals and hydrogen-bonded. Amorphous solids can be classified in an analogous manner (Adler, 1971). In the present context, metals are not of much interest. Due to their low melting points, Van der Waals and hydrogen-bonded solids have not been studied in any significant details. The classification of amorphous semiconductors is therefore essentially confined to ionic and covalent materials. Ionic amorphous semiconductors include the halide and oxide glasses; other ionic amorphous semiconductors which have been investigated are vanadium phosphate and iron phosphate glasses (see Adler, 1971).

Covalent amorphous semiconductors can be subdivided into elemental amorphous materials and other covalent materials which include binary materials. Silicon, germanium, tellurium and selenium belong to the first category. The second category of materials includes As_2Se_3, GeTe as well as boride, arsenide and chalcogenide glasses.

9.1.4 *Thermal conductivity-variation with temperature*

As regards thermal conductivity studies, amorphous semiconductors can be described along with other non-metallic amorphous materials. The thermal conductivity of amorphous materials is significantly lower than that of the crystals. For example, the thermal conductivity of amorphous selenium

(Fig. 9.1) is about two orders of magnitude lower than that of the crystalline material. In the crystalline state, thermal conductivity (λ) depends strongly on the material and any deviation from perfect periodicity brings about significant reduction in λ, particularly in the region of thermal conductivity maximum.

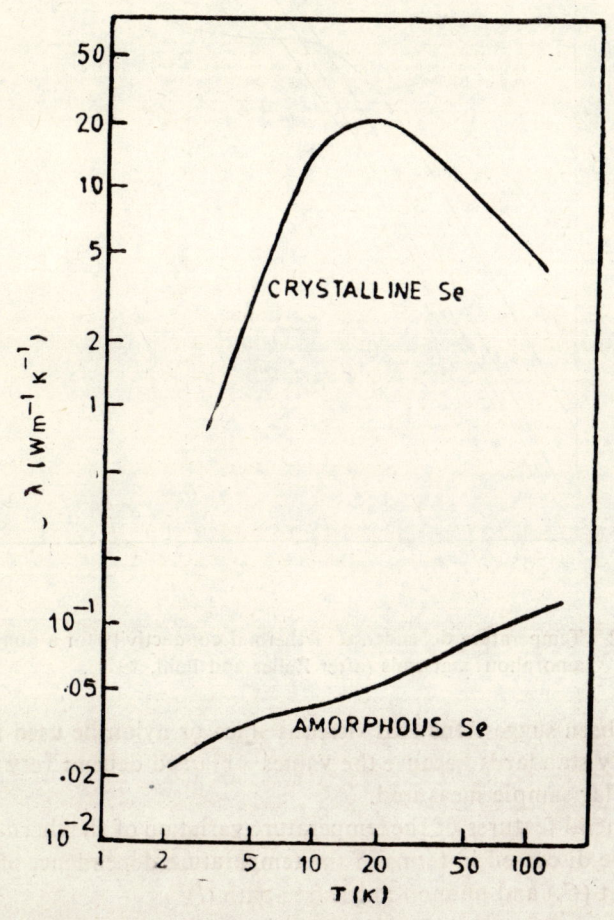

Fig. 9.1 Thermal conductivity of crystalline and amorphous selenium (after White *et al*, 1958).

The difference between the thermal conductivities of various amorphous solids is much less compared to that between different crystals. They all have the same thermal conductivity within a narrow range and the λ-T curves show a plateau near 10 K (Fig. 9.2). Below about 1 K the thermal conductivity varies at T^y where $y = 1 \cdot 8$-2.0. The relative insensitivity of λ to changes in composition points towards the fact that the scattering of phonons must have an origin which is independent of the structural details. Consequently,

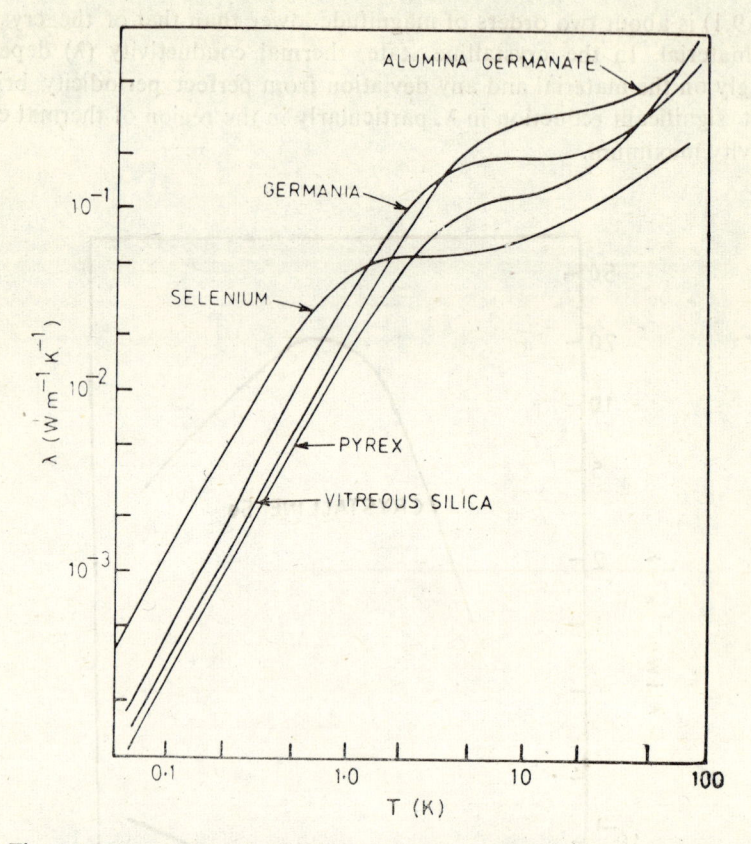

Fig. 9.2 Temperature dependence of thermal conductivity for a number of amorphous materials (after Zeller and Pohl, 1971).

there have been suggestions that vitreous silica or nylon be used as thermal conductivity standards because the values obtained depend very little upon the particular sample measured.

The general features of the temperature variation of the thermal conductivity can be discussed in terms of the temperature dependence of the lattice specific heat (C_v) and phonon mean-free-path (l)

$$\lambda = \tfrac{1}{3} C_v l v_s \tag{9.1}$$

where v_s is the mean phonon (sound) velocity.

At high temperatures in crystalline solids the phonon mean-free-path is limited to the order of a lattice constant and this is also true of amorphous solids. As the specific heats at these temperatures are not very different in the crystalline and amorphous states, the thermal conductivities are also expected to be quite similar. As the temperature decreases the thermal conductivity of a crystal increases following approximately the T^{-1} law. This is due to a rapid increase in the phonon mean-free-path which is reasonably well understood in terms of three-phonon processes (Chap. V). The thermal conductivity of a crystal reaches a maximum and then starts decreasing at lower temperatures, following the temperature variation of the specific heat, as the phonon mean-free-path reaches a value limited by the crystal boundaries.

The thermal conductivity of an amorphous solid, on the other hand, follows a different pattern. Starting with not too different thermal conductivities at high temperatures, the ratio $\lambda_{crystal}/\lambda_{amorph}$ increases as λ_{amorph} decreases with decreasing temperature. In this temperature range, the increase of phonon mean-free-path with decreasing temperature is not as rapid as in the crystal and, therefore, the thermal conductivity follows the specific heat and decreases with decreasing temperature.

Figure 9.3 shows the temperature variation of the phonon mean-free-path in crystalline and vitreous silica. The ratio $l_{crystal}/l_{amorphous}$ has a value around 3 at 500 K. At 10 K this increases to around 14,000 and then decreases with decreasing temperature to a value of 25 at 0.1 K. Above 500 K the thermal conductivity of silica increases rapidly which can be explained on the basis of the heat flow by radiation (Carwile and Hoge, 1966).

9.1.5 *Plateau in the λ–T Curves*

Different mechanisms have been proposed to explain the plateau in the thermal conductivity versus temperature curves around 10 K and the ω^{-4} dependence of the mean-free-path in this region. Klemens (1951) proposed a model according to which most of heat at low temperatures is carried by longitudinal phonons which are assumed to be less effectively scattered by structural irregularities than transverse phonons. The onset of anharmonic coupling of phonons of different branches around 10 K lowers the mean-free-path and gives rise to a plateau.

Klemens (1965) also suggested a resonant scattering of phonons by localized phonons. This could lead to a plateau in a way similar to the formation of the dips in the thermal-conductivity-versus-temperature curves for crystals containing substitutional impurities (Chap. V). Dreyfus *et al.* (1968) suggested a resonant scattering based upon an additional band of localized modes with frequencies above a characteristic frequency ω_R. This band is assumed to damp the phonons with frequencies $\omega > \omega_R$, so that the phonon mean-free-path in this region is a constant, whereas for $\omega < \omega_R$, a dependence, $l \sim \omega^{-2}$ is assumed. This model has been used to explain the plateau in the thermal conductivity as well as the excess specific heat of Se (Lasjaunias and Maynard, 1971) and GeO_2 (Blanc *et al.*, 1971). Leadbetter (1972) proposed an explanation based on the dispersion of the transverse-acoustic branches in amorphous materials.

Zeller and Pohl (1971) proposed an explanation which reduces the problem to that of an isotopic Rayleigh scattering process. In their model every atom is assumed to be displaced from a regular array and the scattering source is represented by a vacancy on each site. They showed that such a process leads to the required frequency dependence of l and the plateau is

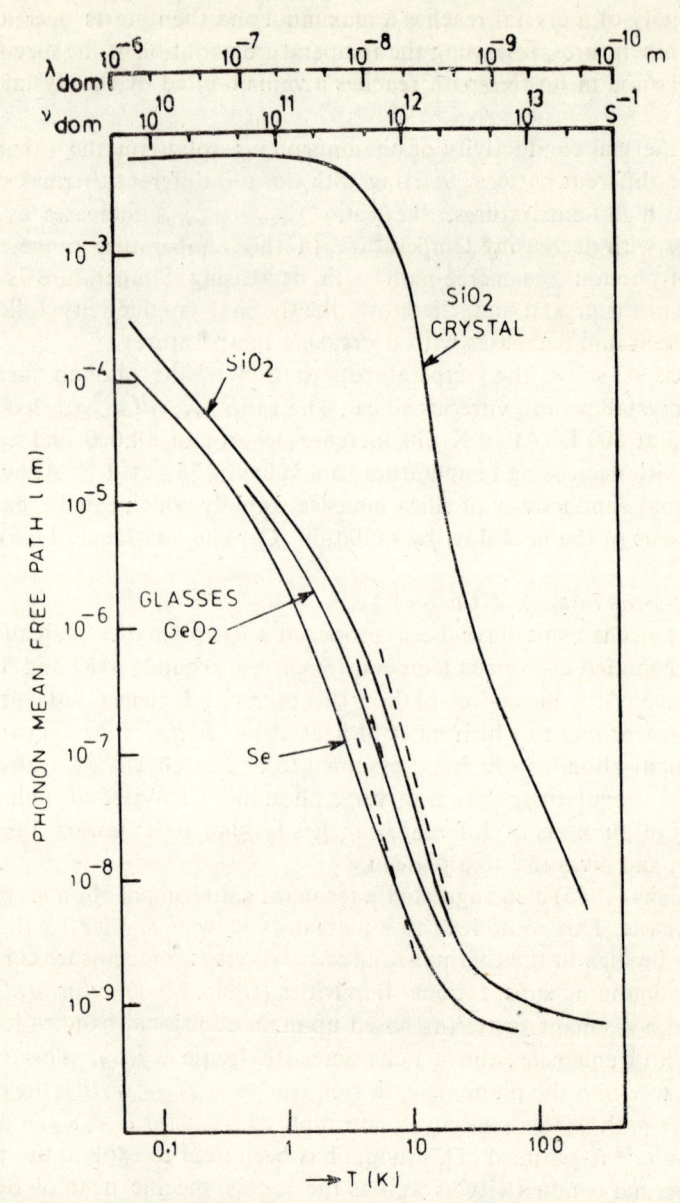

Fig. 9.3 Variation of the phonon mean-free-path with temperature (after Zeller and Pohl, 1971). Dashed lines computed on the mass-difference scattering model. ν_{dom} and λ_{dom} are the frequency and wavelength of dominant phonons.

explained by a cancellation of the temperature dependence of l and C_v. The mean-free-paths were calculated for the dominant phonons as a function of temperature. The results of these calculations show a fairly good agreement with those obtained from the thermal conductivity.

9.1.6 Interpretation of the thermal anomaly at very low temperatures

At very low temperatures (around 1 K and below) the specific heat does not follow the usual T^3 law. The measured specific heat is higher than that expected from the Debye theory and its temperature dependence is much lower than T^3. The specific heat in this region is often fitted to the formula

$$C = C_1 T + C_3 T^3 \tag{9.2}$$

C_3 is generally found to be greater than that computed in the Debye approximation. The linear term is dominant below ~ 0.2 K and does not vary much between different amorphous materials.

Of the various models employed to explain the linear term in the specific heat, the one proposed by Anderson *et al* (1972) has been widely used to obtain comparisons between theory and experiment. According to this model, atoms (or groups of atoms) can tunnel between equilibrium positions at almost the same energy with a smooth distribution of levels between them. This model can yield a specific heat which varies linearly with temperature. Anderson *et al.* further explained the observed T^2 variation of the thermal conductivity by a resonant scattering of phonons from these low-level tunneling systems. A similar model has been developed independently by Phillips (1972). Certain aspects of this theory will be described in the next section.

Other theoretical models describe the effect of structural disorder on the temperature variation of thermal conductivity. Morgan and Smith (1974) introduced the concept of long-range and short-range ordering in amorphous solids. There have been attempts to explain the conductivity in terms of fluctuations in density (Walton, 1974). Joshi (1979) discussed a model in which the thermal conductivity of an amorphous solid has been analysed in terms of the scattering by low-lying tunneling states and by fluctuations in density and elasticity.

9.1.7 The tunneling site model

Tunneling sites have been studied at low temperatures and originate when atoms or molecules alter position or orientation due to quantum-mechanical tunneling (Phillips, 1981, and Bhattacharyya, 1981). This results in a ground state and one (or more) excited states(s). In crystalline material the energies of these excitations are defined in a narrow range since the barriers through which tunneling occurs are similar. In amorphous materials a broad statistical distribution of barriers is expected which results in a broad spectrum of energies and of relaxation times. Moreover, the distribution must depend weakly on the energy E (for small E) so as to fit the T^2 dependence of thermal conductivity and the T-dependence of the excess specific heat. This model has been successful in explaining the results of acoustic measurements on glasses (Jackle, 1972, Jackle *et al*, 1976, Hunklinger and Arnold, 1976, and Golding and Graebner, 1976). The phonon frequencies involved in these measurements are in the range 10^7–10^{10} Hz, which overlap the range of frequencies involved in thermal conductivity measurements. Moreover, the mean-free-paths deducted from acoustic measurements are similar to those obtained from thermal conductivity (Golding *et al.*, 1976 a).

In this model the resonant scattering of phonons by sites of energy E ($= \hbar\omega$) produces a mean-free-path given by (Jackle, 1972, and Jackle *et al*, 1976)

$$l = \left(\frac{Ak_B}{\hbar\omega}\right) \cot h \left(\frac{\hbar\omega}{2k_BT}\right) \qquad (9.3)$$

This explains the temperature-dependence of thermal conductivity but there still remains the difficulty in explaining the plateau (Anderson, 1981). An additional scattering which is associated with tunneling sites (a nonresonant relaxation process) and produces a mean-free-path $l = BT^{-3}$, can be considered along with the resonant scattering to obtain the resulting mean-free-path

$$l^{-1} = [(Ak_B/\hbar\omega) \cot h(\hbar\omega/2k_BT)]^{-1} + [BT^{-3}]^{-1} \qquad (9.4)$$

The nonresonant term prevents the unlimited increase in the mean-free-path and removes the difficulty in explaining the plateau.

Several authors have considered a weak energy dependence for the density of localized states (which was earlier taken to be constant, Piche *et al.*, 1974, Pelous and Vacher, 1976, and Golding *et al.*, 1976 b). Using the density-of-states as deduced from acoustic measurements combined with Eq. (9.4) gives a fit to the experimental thermal conductivity data (Zaitlin and Anderson, 1975a, and Maynard, 1976).

An understanding of the various aspects of the thermal conductivity of amorphous materials is still at a phenomenological level; tunneling sites have not been identified in any amorphous system studied to date. Recent measurements of thermal conductivity by Graebner and Allen (1984) were analysed in terms of the scattering of phonons from two-level tunneling systems, cylindrical inhomogeneities and boundaries. The two-level tunneling systems have been associated with the low-density regions of the material.

9.1.8 *Minimum of thermal conductivity*

The concept of a minimum of thermal conductivity has been discussed in Sec. 6.11. The theoretically calculated minimum was found to be considerably smaller than the measured thermal conductivity of heavily doped semiconducting materials such as silicon–germanium alloys and alloys of lead telluride and cadmium telluride. This leads to a speculation that there is scope for a further reduction in the thermal conductivity of these materials. This could be of interest to researchers looking for improved thermoelectric materials (Slack, 1979, and Rowe and Bhandari, 1983).

Kittel (1949) pointed out that the thermal conductivity of glass should be similar to that of a material whose mean-free-path is of the order of a few interatomic spacings. In that case there does not appear to be much scope for any further reduction in the thermal conductivity. It is of some interest to compare the measured thermal conductivity of a glassy material with the calculated minimum. Figure 9.4 gives a plot of the thermal con-

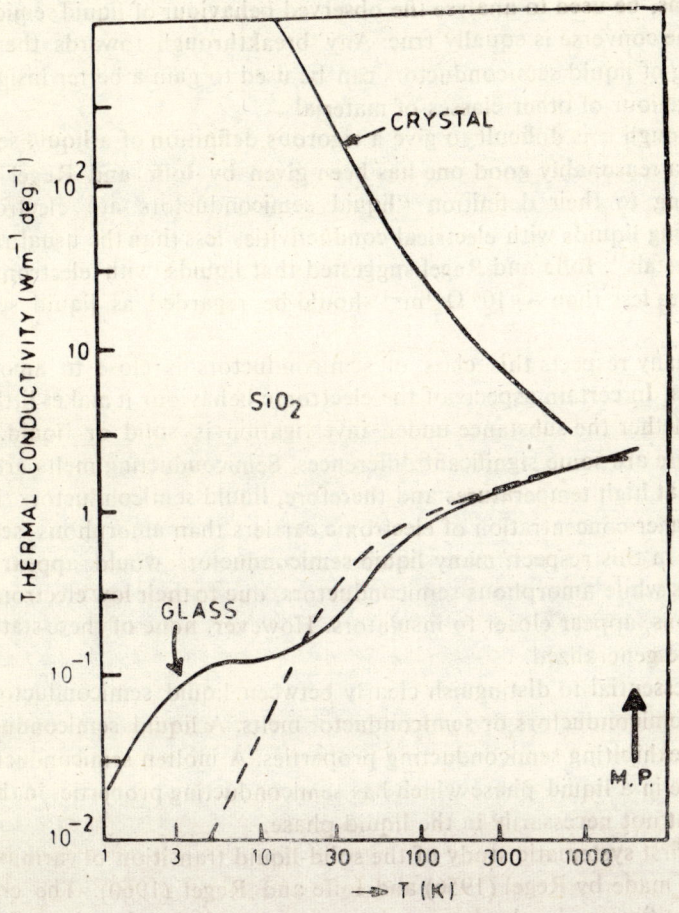

Fig. 9.4 Thermal conductivity versus temperature for α—SiO₂ and crystalline SiO₂. The dashed curve shows the calculated minimum thermal conductivity (after Slack, 1979).

ductivity against temperature for crystalline SiO_2, SiO_2-glass and the calculated minimum. There is a remarkably good agreement between the measured values for SiO_2-glass and the calculated minimum where the photon contribution to thermal conductivity has been subtracted from the measured values. The crystalline material shows a thermal conductivity close to λ_{min} only when the temperature approaches the melting point.

9.2 Liquid semiconductors

9.2.1 Introduction
Liquid semiconductors can be placed in an intermediate position between

several other classes of materials. They have characteristics which partly resemble molten salts, molecular liquids, amorphous solids and liquid metals (Cutler, 1977). An understanding of these material classes could, in certain situations, be used to analyse the observed behaviour of liquid semiconductors. The converse is equally true. Any breakthrough towards the understanding of liquid semiconductors can be used to gain a better insight into the behaviour of other classes of materials.

Although it is difficult to give a rigorous definition of a liquid semiconductor, a reasonably good one has been given by Ioffe and Regel (1960). According to their definition "liquid semiconductors are electronically conducting liquids with electrical conductivities less than the usual range of liquid metals". Ioffe and Regel suggested that liquids with electronic conductivities less than $\sim 10^6\ \Omega^{-1}\mathrm{m}^{-1}$ should be regarded as liquid semiconductors.

In many respects this class of semiconductors is close to amorphous materials. In certain aspects of the electronic behaviour it makes little difference whether the substance under investigation is solid or liquid. However, there are some significant differences. Semiconducting melts are possible only at high temperatures and therefore, liquid semiconductors have a much higher concentration of electronic carriers than amorphous semiconductors. In this respect, many liquid semiconductors would appear closer to metals while amorphous semiconductors, due to their low electronic concentrations, appear closer to insulators. However, none of these statements should be generalized.

It is essential to distinguish clearly between liquid semiconductors and molten semiconductors or semiconductor melts. A liquid semiconductor is a liquid exhibiting semiconducting properties. A molten semiconductor is a substance in a liquid phase which has semiconducting properties in the solid phase but not necessarily in the liquid phase.

The first systematic study of the solid-liquid transition of various materials was made by Regel (1956) and Ioffe and Regel (1960). The criterion for a classification in the liquid phase was based on the sign of $d\sigma/dT$, where σ is the electrical conductivity. A positive value of $d\sigma/dT$ in the liquid state refers to the melt possessing semiconductor properties; a negative sign refers to a metallic melt. All investigated materials were placed in either of the three groups: semicondutor-semiconductor, semiconductor-metal and metal-metal. Regel and Ioffe placed Bi_2S_3, Sb_2S_3, Cu_2S, $CdTe$, $ZnTe$, V_2O_5, Tl_2S, etc. in the first group which retained semiconducting properties in the molten state. Ge, Si and III-V compounds are placed in the second group whereas metals are placed in the third.

A large number of liquid semiconductors are alloys which contain two or more elements. As liquids can possess wide ranges of stoichiometry, liquid semiconductors may be regarded as alloy systems in which the composition varies over a continuous range. Binary systems such as Se-Te, Tl-Te, Tl-Se, Ag-Te, Mg-Bi and V-O can be included in the list of liquid semiconductors. More complicated systems, such as Ga_2Te_3–Ga_2Se_3 have also been investigated.

The electrical and thermal properties of binary systems have been studied as a function of composition and temperature. A study of the electrical conductivity of the system Tl_xTe_{1-x} as a function of x and T shows that for $x > 2/3$, $d\sigma/dT$ is small and positive and σ changes rapidly with x. For $x < 2/3$, $d\sigma/dT$ is large and variation of σ with x is less rapid. A plot of electrical resistivity with x for a fixed temperature shows a maximum at $x = 2/3$ (Cutler, 1971). A comprehensive review of the various properties of liquid semiconductors has been given in texts by Glazov et al. (1969) and Cutler (1977).

9.2.2 Analysis of experimental data

Methods for the measurement of the thermal conductivity of liquid semiconductors have been described in Section 2.4. Accurate measurements at high temperatures are difficult and various measurements on different samples of the same liquid may differ by around 20 per cent from each other. Regel et al (1971) have presented an exhaustive list of the liquids for which thermal conductivities have been measured.

The total thermal conductivity of a melt can be written as

$$\lambda = \lambda_a + \lambda_E + \lambda_{photon} \qquad (9.5)$$

λ_a, which is analogous to the lattice thermal conductivity in solids, is referred to as atomic or molecular thermal conductivity in liquids, λ_E is the contribution to the thermal conductivity from electron processes of heat transfer and λ_{photon} is the photon thermal conductivity.

Atomic or molecular thermal conductivity

The atomic thermal conductivity λ_a has a value of 0.05–0.20 cal m^{-1}s^{-1} deg^{-1} for most of the liquids and sets a lower limit to the thermal conductivity. For most of the known liquid semiconductors, λ_a is very close to a value of 0.1 cal m^{-1}s^{-1} deg^{-1}. This fact can be understood on the following lines: thermal conductivity is related to thermal diffusivity (λ') and the specific heat per unit volume (C) by

$$\lambda_a = \lambda' C \qquad (9.6)$$

Near its melting point the thermal motion of a liquid is largely vibrational in character and according to the law of equipartition of energy, the specific heat is given by $C = 3k_B/(4\pi a^3/3)$; a is the atomic radius. At these high temperatures the vibrational energy diffuses with a mean distance $= 2a$, so that $\lambda' = 4a^2\nu$, where ν is the vibrational frequency of the molecule or atom. With the help of these relations the atomic thermal conductivity is written as

$$\lambda_a = \frac{9}{\pi}\frac{k_B \nu}{a} \qquad (9.7)$$

186 THERMAL CONDUCTION IN SEMICONDUCTORS

For heavier elements 'a' does not vary far from 2 Å and ν is close to 3×10^{12} s^{-1}, and therefore λ_a varies in a narrow range. Figure 9.5 gives a plot of λ_a

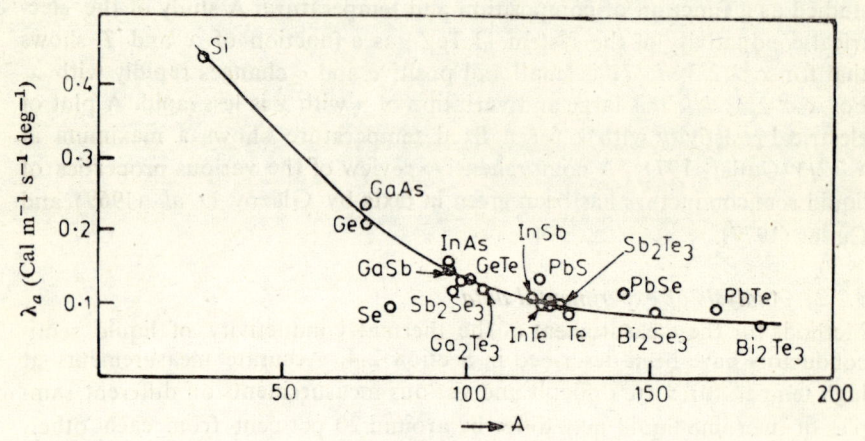

Fig. 9.5 The calculated atomic thermal conductivity at the melting point for various liquids as a function of the average atomic mass (after Regel et al·, 1971).

as a function of average atomic (or molecular) mass at the melting point for various liquids. One of the methods to calculate λ_a is based on Rao's transformed formula (Turnbull, 1961). This gives

$$\lambda_a = G \left[\frac{T_{mp}}{\dfrac{M}{n} \left(\dfrac{V}{n}\right)^{4/3}} \right]^{1/2} \qquad (9.8)$$

n is the number of discrete ions and V the atomic volume. G acquires a constant value for dissociated liquids. This expression is analogous to the formula obtained by Keyes (1959) for solids in the vicinity of the melting point.

In addition to the atomic thermal conductivity which arises from the vibrational motion of the atoms there may exist a contribution to thermal conductivity which arises from the flow of ions or molecules in the direction of the temperature gradient (Regel et al, 1971). This contribution is referred to as the diffusion thermal conductivity and is usually small.

Electronic processes of heat transfer
The term λ_E in Eq. 9.3 can be written as

$$\lambda_E = \lambda_e + \lambda_b + \lambda_{res} \qquad (9.9)$$

λ_e is the electronic (polar) contribution to the thermal conductivity and, for strongly degenerate systems, is related to the electrical conductivity by the Wiedemann–Franz law, $\lambda_e = L\sigma T$, where L is the Lorenz number ($L = (k_B/e)^2 \mathcal{L}$). The limiting values of $\mathcal{L}$ are given by (see **Chap. IV**)

$$\mathcal{L} = (s + 5/2) \qquad -\xi \gg 1$$
$$= \pi^2/3 \qquad \xi \gg 1 \qquad (9.10)$$

s is the scattering parameter and ξ is the reduced Fermi energy (Chap. IV).

λ_b is the bipolar contribution to thermal conductivity and λ_{res} arises due to resonance transfer of ionization energy of the impurity states (Regel et al, 1971). A theoretical value of λ_{res} was obtained by Koshino and Ando (1960 and 1961). This contribution, although significant in the intrinsic range, is difficult to calculate theoretically as the information about various parameters that appear in the formulae are not quite well known in the liquid state (Regel et al, 1971).

A good test of the theory is to plot λ against σT and look for a straight line with a slope equal to L. Perron (1970) plotted λ versus σT for a number of Se−Te alloys (Fig. 9.6) and obtained $L = 2.0 \times 10^{-8}\ V^2 \deg^{-2}$. This

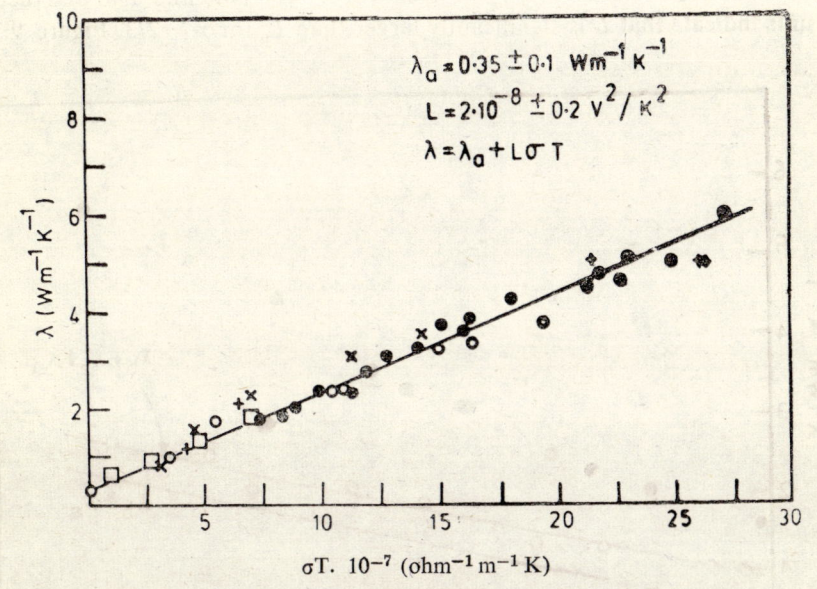

Fig. 9.6 Thermal conductivity versus σT in $Se_x Te_{1-x}$ alloys for $x=0$, 0.05, 0.1, 0.2, 0.3 and 0.4 (after Perron, 1970).

is in good agreement with the theoretical value of $\dfrac{\pi^2}{3}(k_B/e)^2$. For liquids with low electrical conductivity the calculation of electronic contribution to thermal conductivity is somewhat complicated (Cutler, 1977).

In a number of semiconductor melts, L may vary with temperature. It may have a value smaller than or greater than L_0, the limiting value of L in the degenerate limit (Regel et al, 1971). L may have a value lower than L_0 due to the inelastic scattering of the current carriers at low temperatures. L can exceed L_0 in cases of band overlapping due to interband carrier interaction.

The thermal conductivity of antimony-doped antimony selenide and sulphide melts was investigated by Sokolov (1984). A plot of the measured thermal conductivity versus σT for the alloys and for pure Sb_2Se_3 made it possible to obtain the lattice component by extrapolation of the λ–σT curve to low electrical conductivities. A plot of L/L_0 against σT showed a rapid fall with an increase in the electrical conductivity. An important feature of the Sb–Sb_2Te_3 alloys is the small excess of L over L_0 for low values of σT.

Bipolar thermal conductivity

The existence of the forbidden energy gap in the energy spectrum of a semiconductor leads to the bipolar contribution to the thermal conductivity (see Chap. IV). Various aspects of the problem of obtaining estimates of the band gap and λ_b are discussed by Cutler (1977). The case of Tl–Te alloys was discussed by Fedorov and Machuev (1970 a, b). They measured the thermal conductivity of Tl_xTe_{1-x} system for $x = 2/3$, 0.5 and 0.40 and their results indicate that L is significantly larger than L_0 for $x = 2/3$. Figure 9.7

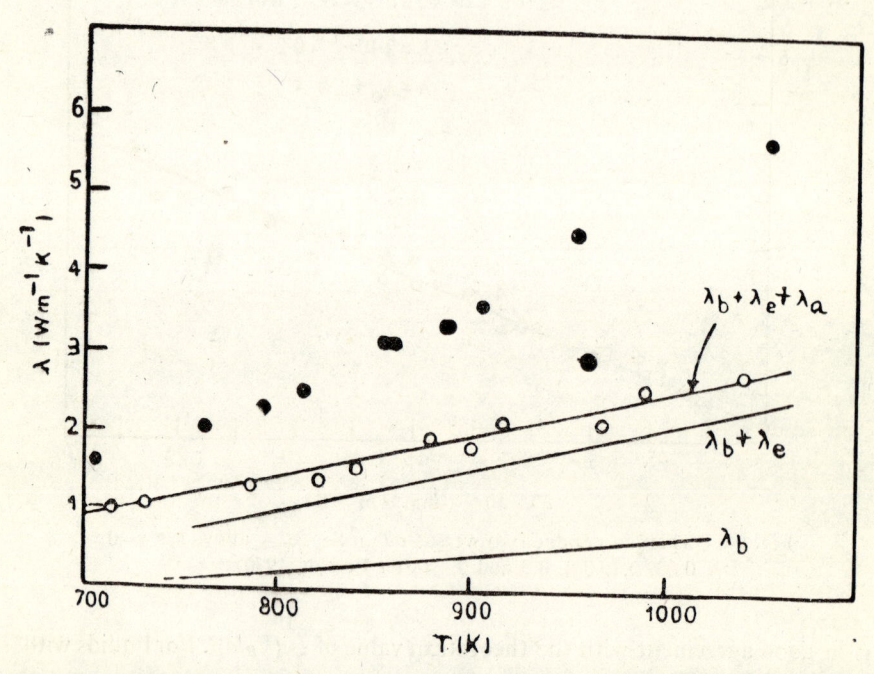

Fig. 9.7 Theoretical curves for λ_b, $\lambda_E(=\lambda_b+\lambda_p)$ and $\lambda(=\lambda_E+\lambda_a)$ plotted against temperature for $Tl_{0.50}Te_{0.50}$ alloys; o, data from Fedorov and Machuev (1970b). ● from Mallon and Cutler (1965) (after Culter, 1977).

shows a plot of λ_b, $\lambda_b + \lambda_e$ and $\lambda_b + \lambda_e + \lambda_a(=\lambda)$ for $x = 0.50$. The atomic thermal conductivity is taken to be equal to 0.1 cal m^{-1} s^{-1} deg^{-1} and the agreement between theoretical and experimental results appears to be good. The results of Mallon and Cutler (1965) donot agree with these results and the authors have attributed this to a systematic error in their measurements (Cutler, 1977).

9.2.3 Concluding remarks

Earlier, during the discussion on glassy materials, it was pointed out that the thermal conductivity of SiO_2-glass is close to λ_{min} over a wide range of temperatures. The disorder in a liquid and a glass appear to be similar as both the systems lack a long-range order. However, a study of the available information on the thermal conductivity of liquids shows that thermal conductivity of pure liquids is not as close to the minimum thermal conductivity as that of glasses. The study of the thermal conductivity of liquid mixtures offers some scope for further reduction (Slack, 1979).

Investigations of the thermal properties of liquid semiconductors have practical applications, particularly in developing thermoelectric devices. Some substances like complex copper chalcogenides have fairly high values of thermoelectric power in the liquid state. The dimensionless thermoelectric figure-of-merit $ZT(=\alpha^2\sigma T/\lambda)$ acquires a value $\simeq 1$ in the best known thermoelectric materials and the values of ZT reported for many liquid semiconductors are not much different. As large temperature gradients can be established in liquids, thermoelectric conversion efficiency can, in principle, be increased by the use of these materials as they can use more efficiently the heat evolved in nuclear reactors or that collected by solar concentrators (Cadoff and Miller, 1960, and Garner, 1963).

References

Adler, J. (1971), *Amorphous Semiconductors*, CRC Press, Cleveland, Ohio.
Anderson, P.W. (1958), *Phys. Rev.* **109**, 1492.
Anderson, P.W., Halperin, B.I. and Varma, C.M. (1972), *Phil. Mag.* **25**, 1.
Anderson, A.C. (1981), *Topics in Current Physics* (ed. W.A. Phillips), Springer Verlag, Vol. 24, p. 65.
Bell R.J. (1972), *Repts Progr. Phys.* **35**, 1315,
Bhattacharyya, A. (1981), *Contemp. Phys.* **22**, 117.
Bottger, H. (1974), *Phys. Stat. Solidi (b)* **62**, 9.
Blanc, J., Brochter, D., Lasjaunias, J., Maynard, R. and Ribeyro, A. (1971), *Proc. 12th Int. Conf. on Low Temp. Physics*, Kyoto (ed. E. Kanda), Acad. Press of Japan, Kyoto, 1971.
Cadoff, I.B. and Miller, E. (1960), *Thermoelectric Materials and Devices*, Reinhold, New York.
Carwile, H.C.K. and Hoge, H J. (1966), *Tech. Rep. No. 67-7-PR*, U.S. Army. Natick Labs
Cohen, M.H., Fritzsche, H. and Ovshinsky, S.R. (1969), *Phys. Rev. Lett* **22**, 1065.
Cohen, M.H. and Lucovsky, G. (1972), (ed.) *Amorphous and Liquid Semiconductors*, *Proc. 4th Int. Conf*, Ann Arbor, Michigan (1971), North Holland, Amsterdam.
Cutler, M. (1971), *Phil. Mag.* **24**, 381.

Cutler, M. (1977), *Liquid Semiconductors*, Acad. Press, New York.
Dean, P. (1972), *Rev. Mod. Phys.* 44, 127.
Dreyfus, B., Ffernandes, N.C. and Maynard, R. (1968), *Phys. Lett.* A26, 647.
Fedorov, V.I. and Machuev, V.I. (1970 a), *Sov. Phys.-Solid St.* 12, 221.
Fedorov, V.I. and Machuev, V.I. (1970 b), *Sov. Phys.-Solid St.* 12, 484.
Garner, J.W. (1963), *Engl. Electric J* 18, 16.
Glazov, V.M., Chizhevskaya, S.N. and Glagzoleva, N.N. (1969), *Liquid Semiconductors*, Plenum Press, New York, (Translation A. Tybelewicz).
Golding, B. and Graebner, J.E. (1976), *Phys. Rev. Lett.* 37, 852.
Golding, B., Graebner, J.E. and Schultz, R.J. (1976 a), *Phys. Rev.* B14, 1660.
Golding, B., Graebner, J.E. and Kane, A.B. (1976 b), *Phys. Rev. Lett.* 37, 1240.
Graebner, J.E. and Allen, L.C. (1984), *Phys. Rev.* B29, 5626.
Gubanov, A.I. (1965), *Quantum Theory of Amorphous Conductors*, Consultants Bureau, New York.
Hunklinger, G. and Arnold, W. (1976), *Physical Acoustics* (ed. W.P. Mason and Thurston), Acad. Press, Vol. 12, 155.
Hori, J. (1968), *Spectral Properties of Disordered Chains and Lattices*, Pergamon Press, Oxford.
Ioffe, A.F. and Regel, A.R. (1960), *Progr. Semicond.* 4, 237.
Jackle, J. (1972), *Z. Phys.* 257, 212.
Jackle, J., Piche, L., Arnold, W. and Hunklinger, W. (1976), *J. Non Cryst. Solids* 20, 365.
Joshi, Y.P. (1979), *Phys. Stat. Solidi* (b) 95, 317.
Keyes, R.W. (1959), *Phys. Rev.* 115, 564.
Kittel, C. (1949), *Phys. Rev.* 75, 972.
Klemens, P.G. (1951), *Proc. Roy. Soc.* A208, 108.
Klemens, P.G. (1965), *Physics of Non Cryst. Solids* (ed. J.A. Prins), North Holland, Amsterdam, p 162.
Koshino, S. and Ando, T. (1960), *J. Phys. Soc. Japan* 15, 1538.
Koshino, S. and Ando, T. (1961), *J. Phys. Soc. Japan* 16, 1151.
Lasjaunias, J.C. and Maynard, R. (1971), *J. Non Cryst Sol.* 6, 101.
Leadbetter, A.J. (1972), *Proc. Int. Conf. on Phonon Scattering in Solids* (ed. A.J. Albany), La Doc. Ir., Paris, p 338.
Mallon, C.E. and Cutler, H. (1965), *Phil. Mag.* 11, 667.
Maynard, R. (1976), *Phonon Scattering in Solids* (ed. L.J. Challis, V.V. Rampton and A.F.G. Wyatt), Plenum, New York.
Morgan, G.J. and Smith, D. (1974), *J. Phys. C: Sol. St. Phys.* 7, 649.
Mott, N.F. (1967), *Adv. Phys.* 16, 49.
Mott, N.F. and Davis, E.A. (1971), *Electronic Processes in Noncrystalline Materials*, Clarendon Press, Oxford (ed. II).
Pelous, J. and Vacher, R. (1976), *Solid St. Comm.* 19, 627.
Perron, J.C. (1970), *Rev. Phys. Appl.* 5, 611.
Phillips, W.A. (1972), *J. Low Temp. Phys.* 7, 351.
Phillips, W.A. (1981), *Topics in Current Physics* (ed. W.A. Phillips), Springer Verlag, Vol 24.
Piche, L., Maynard, R., Hunklinger, S. and Jackle, J. (1974), *Phys. Rev. Lett* 32, 1426.

Pohl, R.O. (1975), Phonon Scattering in Amorphous Solids, *Int. Conf. on Phonon Scatt. in Solids*, University of Nottingham, p. 107.

Rao, M.R. (1941), *Phys. Rev. 59*, 212.

Raychaudhury, A.K., Peech, J.M. and Pohl, R.O. (1979), Phonon Scatt. in Glasses and Highly Disordered Cryst., *Int. Conf. on Phonon Scatt. in Matter*, Brown Univ., p 45.

Regel, A.R , (1956), *Dissertation*, Univ. of Leningrad.

Regel, A.R., Smirnov, I.A. and Shadrichev, E.V. (1971), *Phys. Stat. Solidi 5*, 13.

Rowe, D.M. and Bhandari, C.M. (1983), *Modern Thermoelectrics*, Holt Saunders, London.

Slack, G.A. (1979), *Solid State Physics*, (ed. Ehrenreich *et al.*), Vol. 34, p 1.

Sokolov, V.N. (1984), *Sov. Phys.—Semiconductors 18*, 495

Tauc, J. (1974) (ed.), *Amorphous and Liquid Semiconductors*, Plenum Press, New York.

Turnbull, A.G. (1961), *Austral. J. Appl. Sci. 12*, 324.

Walton, D. (1974), *Solid St. Comm. 14*, 335.

White G.K., Woods, S.B. and Elford, M.T. (1958), *Phys. Rev. 112*, 113.

Zaitlin, M.P. and Anderson, A.C. (1974), *Phys. Rev. Lett. 33*, 1158.

Zaitlin, M.P. and Anderson, A.C. (1975 a), *Phys Stat. Sol. (b) 71*, 323.

Zaitlin, M.P. and Anderson, A.C. (1975 b), *Phys. Rev. B12*, 4475.

Zeller, R.C. and Pohl, R.O. (1971), *Phys. Rev. B4*, 2029.

Chapter X

Miscellaneous Semiconductors

10.1 Introduction

Most of the discussion on the study of thermal transport has been related to crystalline semiconductors where the observed thermal conductivity data could be analysed on the basis of theories of heat conduction by phonons and electrons (or holes) taking into account the various scattering mechanisms which tend to limit their mean-free-paths. Chapter IX was devoted to a class of materials which could not be accomodated within the framework of crystalline materials and, consequently, required separate treatment. The discussion on crystalline and non-crystalline materials covered almost all the topics of interest as regards the analysis of thermal conductivity. However, there are some topics which, although could be included within the main text of this book, warrant a separate identification and treatment. This chapter is devoted to miscellaneous topics: magnetic semiconductors and heterogeneous materials. As these topics are unrelated to each other they are discussed under the title of "miscellaneous semiconductors".

Magnetic semiconductors exhibit a number of interesting features at low temperatures. However, little interest has been expressed in thermal conductivity studies of organic semiconductors; their study under a separate title appears to be justified more on the basis of their electrical properties rather than their thermal conductivity.

10.2 Magnetic semiconductors

10.2.1 *Introduction*
Some metallic oxides behave as semiconductors with relatively large energy gaps. Of these Cu_2O, ZnO and the transition-metal oxides have been thoroughly investigated. In general the metal-ions in these materials are characterized by incomplete d-shells, and on this ground they might be expected to exhibit a degree of metallic behaviour. Some oxides do behave like metals with the electrical resistivity steadily increasing with temperature. Others, such as pure NiO, appear more similar to insulators while

V_2O_3 shows metallic behaviour at high temperatures. At a critical temperature its electrical conductivity drops by a factor of about 10^6 and below this temperature it behaves as a semiconductor.

Various spinels and garnets which are ferrimagnetic exhibit semiconducting properties. Ferromagnetic materials like EuO, EuS and EuSe also exhibit semiconducting properties.

The thermal conductivity of magnetic semiconductors may possess interesting features at very low temperatures. These features are essentially due to their magnetic rather than their semiconducting behaviour. However, they need to be discussed under a separate class of materials although they cannot be distinguished from magnetic insulators as regards their special thermal conductivity features at low temperatures.

In some of these materials heat may be carried by magnons (quantized spin-waves) at very low temperatures, and the experimental thermal conductivity data analysed in terms of the separate contributions to heat transport from phonons and magnons. Magnons in their passage through the material are scattered by crystal boundaries and magnetic impurities. Scattering by other magnons and phonons may also be present. A significant amount of magnon contribution to thermal conductivity has been detected in several materials.

On the other hand, there may be significant changes in the thermal conductivity at the metal-semiconductor transition in materials like V_2O_3. However, the study of thermal conductivity in this context has not attracted a great deal of interest and the discussion on the thermal conductivity of magnetic semiconductors will, in this chapter, be confined mainly to the study of magnon contribution.

10.2.2 *Magnon and phonon contributions to thermal conductivity*

As already stated, heat conduction in magnetic insulators and semiconductors can be adequately described in terms of the separate contributions from phonons and magnons with proper allowance being made for the various interactions of these excitations with one another and with different types of imperfections, impurities and with boundaries.

The first attempt in this direction was made by Rezanov and Cherepanov (1953) who calculated the thermal conductivity of a ferromagnetic metal. In this case, conduction is mainly by electrons, and magnons merely act as scattering centres which reduce the electronic contribution to thermal conductivity. Identification of the magnon contribution in this situation was difficult. A more interesting situation was pointed out by Sato (1955) in analysing the thermal conductivity of a ferromagnetic insulator. He neglected the phonon-magnon interaction and assumed that at sufficiently low temperatures the mean-free-paths of phonons and magnons are limited only by crystal boundaries, are temperature independent and comparable in magnitude. He found that the thermal conductivity due to magnons (λ_m)

varies as T^2 and is greater than the phonon thermal conductivity (λ_{ph}) which varies as T^3.

Douthett and Friedberg (1961) investigated the low-temperature thermal conductivity of ferrite crystals which behave like insulators at sufficiently low temperatures. The dispersion relations for magnons were taken to be similar to that in a ferromagnet. Using a dispersion law (Kittel, 1964) $\hbar\omega_\lambda = DK^2_\lambda + g\mu_B B$, where K_λ refers to the magnon wave vector, g is the spectroscopic splitting factor, μ_B is the Bohr magneton (the second term on the right being the familiar Zeeman term; see Kittel, 1964), they showed that the magnon thermal conductivity in the presence of a magnetic field B is given by

$$\lambda_m(B) = \frac{l_s k_B^3 T^2}{3\pi hD} \sum_n \left\{ \left(\frac{\mu_B g B}{k_B T}\right)^2 \frac{1}{n} + \left(\frac{4g\mu_B B}{k_B T}\right) \frac{1}{n^2} + \frac{6}{n^3} \right\} \exp\left(-ng\mu_B B/k_B T\right) \tag{10.1}$$

For zero-field this reduces to

$$\lambda_m(0) = 0.765 \, l_s k_B^3 T^2/hD \tag{10.2}$$

Here l_s is the magnon mean-free-path and is assumed to be constant. For a constant phonon mean-free path l_p, they calculated the phonon conductivity to be

$$\lambda_{ph}(0) = 326 \, k_B^4 l_p T^3/h^3 v_s^2 \tag{10.3}$$

v_s being the average sound velocity.

If l_p and l_s are taken to be of equal magnitude and $v_s = 5 \times 10^3$ ms^{-1}, and $D = 10^{-39}$ Jm^2, the thermal conductivity at 2K in a field of 2.4 T is expected to be reduced by $\Delta\lambda/\lambda = -0.3$. Friedberg and Harris (1963) studied the thermal conductivity of yttrium iron-garnet in the temperature range 1.3–40 K and an external field varying from 0 to 2.4 T applied parallel to the [100] or [111] axis. The observed reduction in thermal conductivity in the liquid-helium temperature region was found to be of the same order as predicted by the above theory. They further showed that below 6K, the thermal conductivity could be well represented by $\lambda(0) = AT^2 + BT^3$, where the first term represents the contribution due to magnons and the second is the familiar phonon term. The constants identified as $A = 0.765 \, l_s k_B^3/hD$ and $B = 326 \, k_B^4 l_p/h^3 v_s^2$, could be obtained from the extrapolated intercept and the slope of the $\lambda(0)/T^2$ versus T curves. l_p and l_s can be calculated and with $v_s = 5 \times 10^3 ms^{-1}$, $D = 0.83 \times 10^{-39}$ Jm^2, they obtained $l_s = 6.0 \times 10^{-5}$ m and $l_p = 5.8 \times 10^{-5}$m. Further a plot of $\lambda(H)/T^2 - \lambda_m(H)/T^2$ versus T was found to be a straight line with the slope of $\lambda(0)/T^2$ and zero intercept. This suggested that $\lambda(B) - \lambda_m(B) = \lambda_{ph} = \lambda_{ph}(0)$ and that λ_{ph} is unaffected by the field.

Luthi (1962) and Douglass (1963) obtained somewhat similar results. McCollum et al (1964) detected a magnon contribution to thermal conductivity in EuS.

Douglass reported measurements on two single crystal samples of yttrium irong-arnet (YIG) between 0.4 and 20 K and in magnetic fields varying from 0 to 2T. A 70% decrease in thermal conductivity occurred when a strong field was applied at 0.5 K. The reductions were smaller at higher temperatures (Fig. 10.1).

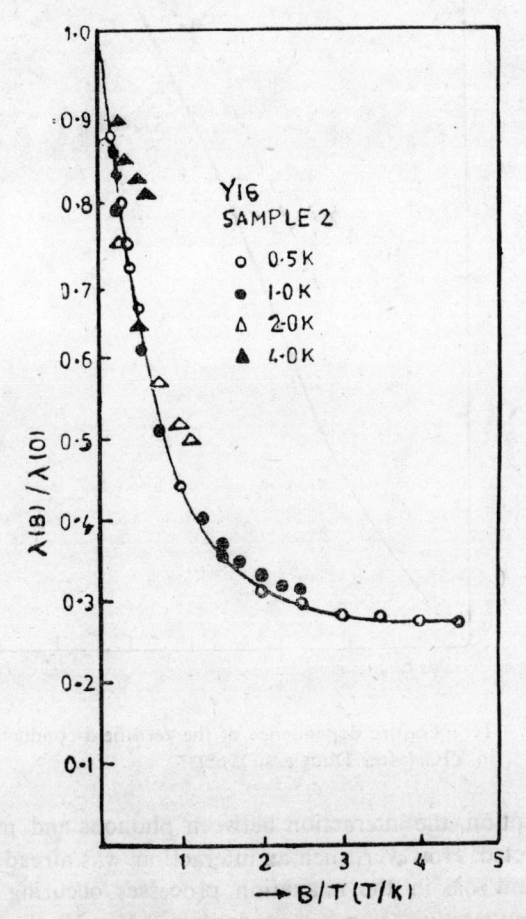

Fig. 10.1 Plot of $\lambda(B)/\lambda(0)$ against B/T, when the magnetic field is parallel to the temperature gradient. The zero-field conductivity was obtained by a linear extrapolation of the curve above 1KOe (After Douglass, 1963).

This reduction in thermal conductivity could be attributed to either: (a) the quenching of the magnon transport by magnetic field, or (b) a decrease of the phonon relaxation length. In Fig. 10.2 is presented the observed thermal conductivity for the two samples as a function of temperature. The difference in the behaviour of the two samples could be attributed to the effect of impurities and other inhomogeneities apart from the crystal boundaries. The zero-field low temperature thermal conductivity of sample 1 is very close to the predicted T^2 dependence.

196 THERMAL CONDUCTION IN SEMICONDUCTORS

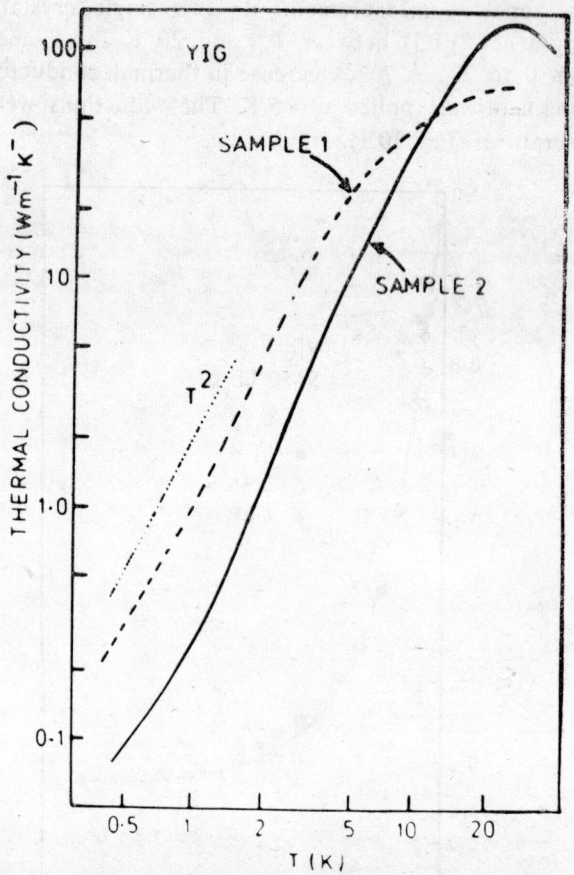

Fig. 10.2 Temperature dependence of the zero-field conductivity in YIG (after Douglass, 1963)

In this description, the interaction between phonons and magnons was completely neglected. However, such an interaction was already known to play an important role in the relaxation processes occuring in magnetic crystals at low temperature (Van Kranendonk and Van Vleck, 1958). Kittel and Abraham (1953), and Kaganov and Tsukernik (1959) studied this problem and calculated the time required for the establishment of equilibrium between the phonon and magnon systems. Sinha and Upadhyaya (1962) developed a microscopic theory of phonon-magnon interaction in magnetic crystals and obtained expressions for phonon-magnon ralaxation times. Bhandari and Verma (1966) presented an analysis of the thermal conductivity data of YIG in the temperature range 0.5–20 K. In this analysis the total thermal conductivity (λ) is written as a sum of separate contributions from phonons (λ_{ph}) and magnons (λ_m)

$$\lambda = \lambda_{ph} + \lambda_m \qquad (10.4)$$

A calculation of magnon thermal conductivity (λ_m) was proposed by

Callaway and Boyd (1964) in a way analogous to the calculation of the phonon conductivity. λ_m is then given by

$$\lambda_m = \frac{k_B}{3(2\pi)^2 h^2} \int \left(\frac{E}{k_B T}\right)^2 \frac{\exp(E/k_B T)}{(\exp(E/k_B T) - 1)^2} (\nabla_k E)^2 \tau(K_\lambda) dK_\lambda \quad (10.5)$$

$E = E(\mathbf{K}_\lambda)$ is the energy of the magnon of wave vector $\mathbf{K}_\lambda$. The effective relaxation time $\tau(\mathbf{k}_\lambda)$ of the magnon can be obtained by adding the inverse relaxation times for various magnon scattering processes. This may include the relaxation times for the scattering by boundaries and by magnetic defects apart from magnon-magnon and magnon-phonon scattering relaxation times. In a real situation it is difficult to include all these scattering rates without introducing a large number of adjustable parameters. Bhandari and Verma (1966) obtained a good fit with the experimental thermal conductivity in YIG in the temperature range 0.5-20K by assuming magnon mean-free-path to be limited by boundaries, magnetic defects and magnon-phonon scattering while for phonons the dominant scattering mechanisms were taken to be the crystal boundaries and point-defects.

Walton *et al* (1973) investigated the magnetic-field dependence of the thermal conductivity of YIG below 1 K and in fields upto 4 T. Assuming the heat flow by a system of non-interacting phonons and magnons, the experimental thermal conductivity data cannot be explained in a satisfactory way. The observed conductivity decreased faster with increasing field than that expected on the basis of non-interacting phonon and magnon modes and it was found that coupled magnetoelastic modes can explain the magnetic-field dependence of the thermal conductivity in a more satisfactory way. The magnetoelastic coupling of magnons and phonons in a ferromagnetic dielectric had earlier been treated by Kittel (1958).

Although the theoretical basis of heat transport by magnons in magnetic insulators has been well established, experimental confirmation of a magnon thermal conductivity has not always been possible. In some substances, such as ferromagnet EuS, ferrimagnet Li-ferrite and antiferromagnet $CoCl_2 \cdot 6H_2O$ and $GdVO_4$, there is evidence for magnon heat transport. However, in substances such as the ferromagnet $GdCl_3$, the ferrimagnet $MnFe_2O_4$ and the antiferromagnets $RbMnF_3$ and MnF_2, no evidence of magnon conduction could be found.

The experimentally measured conductivity depends upon the phonon and magnon intrinsic conductivities λ_p and λ_m and also on a characteristic time τ_{mp}. This is the time in which the magnon temperature comes to equilibrium with the phonon temperature in an isolated system (Sanders and Walton, 1977). In a conventional thermal conductivity experiment some degree of interaction is required between magnons and phonons in order to observe the magnon heat transport. Only phonons are generated at the hot end and absorbed at the cold end. Heat can enter or leave the magnon system only if the coupling between the phonon and magnon systems is strong enough. The cases of YIG and MnF_2 have been discussed by Sanders and

Walton and they have shown that a short τ_{mp} in YIG allows λ_m to be observed while a large value of τ_{mp} in MnF_2 may be responsible for the absence of λ_m in a conventional thermal conductivity experiment.

10.2.3 *Other materials*

The results of thermal conductivity investigations on several other magnetic materials show interesting features. For example, the transition to an ordered magnetic phase in antiferromagnetic CoF_2 is reflected in the thermal conductivity measurements. There is a pronounced minimum in the λ-T curve at the Neel temperature of 38K (Slack, 1961). Unusual features have been reported in the thermal conductivity behaviour of Fe_3O_4 at 119K when λ increases by a factor of 1.6 (Slack, 1962). Below this temperature there is an ordering of the Fe^{2+} and Fe^{3+} ions on the octahedral sites. The thermal conductivities of these materials can be explained by assuming that phonons are the only important carriers of heat. Apart from the usual standard phonon-scattering processes, one additional scattering mechanism appears in these cases; this is the scattering of phonons from local disorder in the lattice formed by the magnetic moments of ferromagnetic ions (Slack, 1962).

The temperature variation of thermal conductivity for several mixed crystals $Mg_{1-x}Fe_xAl_2O_4$ was also investigated by Slack (1964). With an increasing content of Fe, thermal conductivity was found to decrease progressively. Several of these crystals showed a kink around 11 K in their λ-T curves. This was interpreted in terms of the phonon-assisted transition between the several low-lying energy levels of the $3d^6$ configuration, and is not due to any cooperative, long-range magnetic ordering. The scattering from the Fe^{2+}, $3d^6$ configuration in tetrahedral sites involves a series of several equally spaced magnetic levels. This is similar to the case of Fe^{2+} in CdTe discussed in Chap. VIII (Slack and Galginaitis, 1964).

10.3 Organic semiconductors

10.3.1 *Introduction*

Semiconducting properties are exhibited by a number of organic materials. However, these materials are not available in a state of purity approaching that of silicon or germanium. There are difficulties in preparing good quality crystals of appreciable size as these materials are generally soft and fragile.

The study of electronic conduction in organic materials accompanied their use in the sensitization of photographic emulsions and its possible importance in the fundamental physical processes of living organisms (Szent-Gyorgyi, 1946, and Garrett, 1959).

Anthracene is one of the most intensively studied organic semiconductors. Other aromatic hydrocarbons, such as naphthalene and tetracene are amongst the best known organic semiconductors; both are derivatives of benzene and form monoclinic crystals (Gutman and Lyons, 1967).

Various electronic properties of these materials have received considerable attention during recent years. However, to date, there appears to be little interest in the study of their thermal conductivity. Unlike the thermal conductivities of the magnetic materials discussed in the previous section, organic semiconductors do not appear to possess any remarkable features in their thermal conductivity behaviour.

Friedman (1964) has given formulae for various transport coefficients of certain organic semiconductors. Assuming the bandwidth to be smaller than $k_B T$, explicit expressions are obtained for a number of transport coefficients including thermoelectric power and thermal conductivity. The main feature that distinguishes these materials from other conventional semiconductors is their narrow energy bandwidth, specially when $\leqslant k_B T_r$, where T_r is the room temperature.

The expressions derived by Friedman for the electronic thermal conductivity are expressed in terms of relevant intermolecular transfer integrals and the electron relaxation times for various crystallographic directions. The relaxation time depends upon the particular scattering mechanism. Structural defects in the lattice, such as dislocations, impurities and phonons are the main scattering agents in these crystals. The interactions between the electronic and vibrational motions can be of various types depending upon whether the time for which an electron stays on a site is larger or shorter than a vibrational period. Under certain conditions, the interaction can lead to a trapping of the electrons in self-induced potential wells. The entity which now moves along the crystal lattice is an electron accompanied by a localized vibration and is referred to as a polaron. Siebrand (1962 and 1963) applied polaron theory to molecular crystals.

The lattice thermal conductivity of these crystals has not been studied in great details, unlike in the case of conventional semiconductors. It appears that lattice conduction models similar to those applicable to conventional semiconductors could be applied to these crystals as well. The next section describes some of the results of thermal conductivity measurements in organic crystals.

10.3.2 *Thermal conductivity of naphthalene*

Thermal conductivity of polycrystalline naphthalene was measured between liquid nitrogen temperature and its melting point (Uberreiter and Orthmann, 1950, and Mogilevskii and Surin, 1972). The temperature dependence of the thermal resistivity W (defined by $\lambda = 1/W = B'/T$) can be well approximated by a straight line (Fig. 10.3). The constant B', which represents anharmonicity, was found to be 19–20 cal m^{-1}s^{-1}. The value of this constant obtained from theoretical considerations based upon the Leibfried-Schlomann formula (see Chap. VI) yields a value $B' = 3.0$ ($\theta_D = 127$ K and $\gamma = 2.3$), which is much less than the value obtained from the W-T plot. There appears to be some theoretical interest in explaining the magnitude of B'

200 THERMAL CONDUCTION IN SEMICONDUCTORS

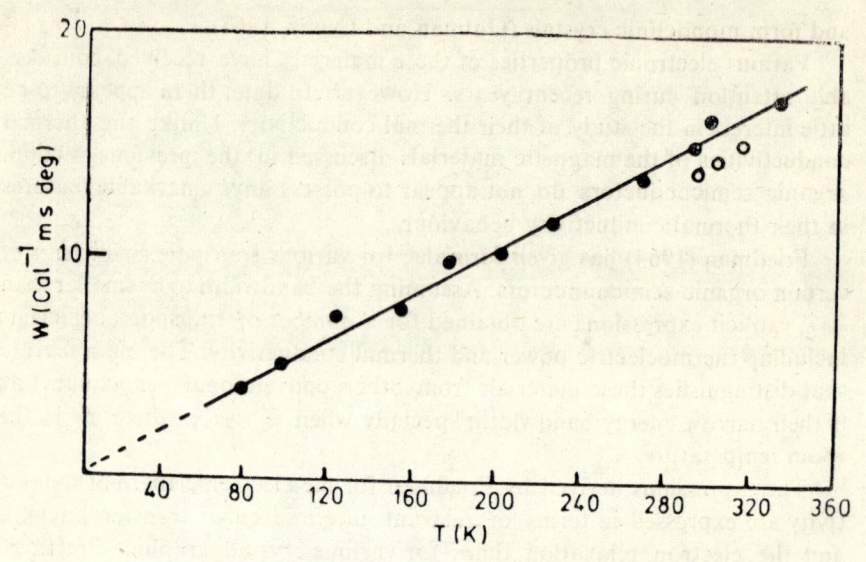

Fig. 10.3 Thermal resistivity versus temperature in naphthalene (after Mogilevskii and Surin, 1972), o-correspond to results of measurements by Uberreiter and Orthman (1950).

although the observed T-dependence of thermal conductivity is very close to the $1/T$ law. In view of the difficulties in obtaining information on the vibrational spectra and the nature of anharmonic interactions in molecular crystals, more detailed analysis of their thermal conductivity behaviour will not be taken up at this stage.

Ross et al (1979 II) reported measurements of the thermal conductivity of naphthalene and anthracene under pressure. Fig. 10.4 and 10.5 show the results of their measurements of the thermal conductivity of these organic semiconductors along with that of benzene including the revised values for its second phase. For a perfect crystal at constant volume theory predicts the usual $1/T$ law while near the melting point a stronger temperature dependence is expected due to the effect of thermal expansion (Klemens, 1977). At a pressure of 0.1 GPa this type of behaviour is observed in benzene I and naphthalene. A weaker dependence of thermal conductivity on temperature than T^{-1} has been attributed to the presence of structural disorder (Ross et al., 1979 I).

10.4 Heterogeneous solids

10.4.1 *Random mixture of two or more phases*
The theories of thermal conductivity of solids described earlier in the book have been based on the assumption that the material is homogeneous. A large number of materials of technological importance are heterogeneous and it is important to try to understand their thermal conductivity behaviour.

MISCELLANEOUS SEMICONDUCTORS 201

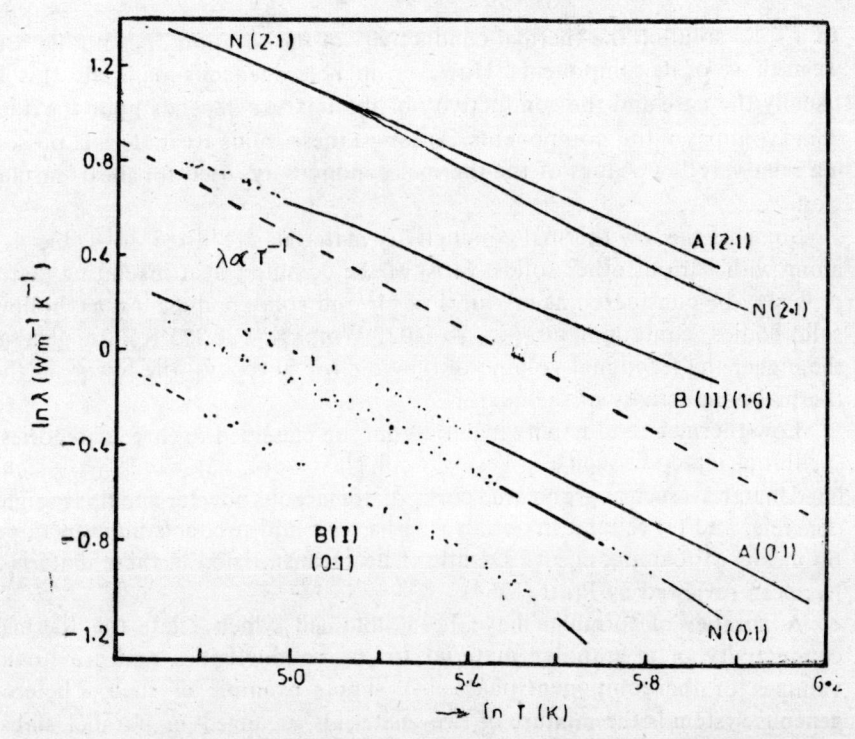

Fig. 10.4 Variation of thermal conductivity with temperature at constant pressure. A, B, N refer to anthracene, benzene and naphthalene. Within the parentheses are given pressures in GPa. The dashed line gives the expected T^{-1} dependence (after Ross et al., 1979).

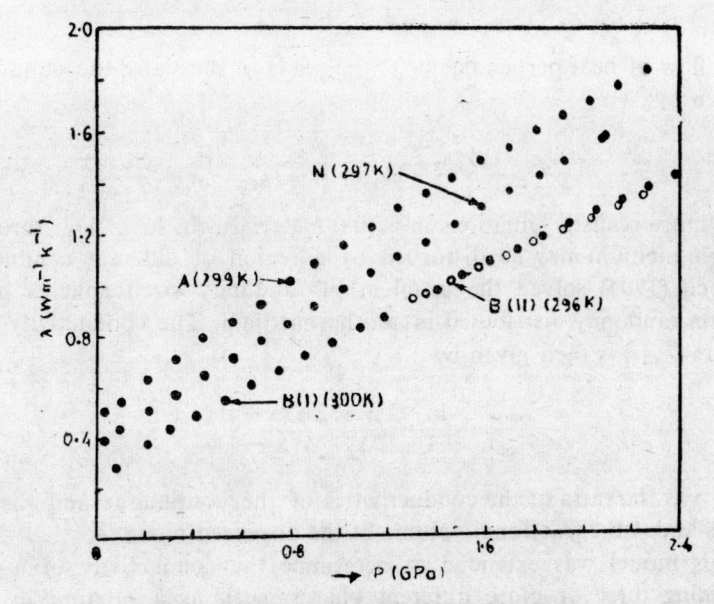

Fig. 10.5 Variation of thermal conductivity with pressure at constant temperature. Benzene I(BI), benzene II(BII), naphthalene (N) and anthracene (A) (after Ross et al, 1979).

In a solid solution the thermal conductivity of the solution does not lie between those of its components. However, in heterogeneous materials this is usually the case and the conductivity of the mixture depends upon the fractional volumes of the components. Amongst these solids are materials, possesing relatively low values of the thermal conductivity, used for heat insulation.

Some of the low thermal conductivity materials are listed in Table 1.1 along with various other solids. Most of the common heat insulating materials may be considered as mixtures of air and solid bodies, or as hollow solid bodies containing air ($\lambda_{air} \sim 0.025$ Wm^{-1} K^{-1} at 273 K). In general the higher the fractional volume of the void (or porosity), the lower is the thermal conductivity of the mixture.

Low-thermal-conductivity materials may be classified in three categories: (a) fibrous materials such as asbestos, wool, glass-wool, slag-wool; (b) granulated materials such as granulated cork, diatomaceous powder and lightweight concrete, and (c) cellular materials such as cork and products manufactured by means of foaming agents. Details of heat transmission in these materials has been reviewed by Pratt (1969).

A number of formulae have been obtained which relate the thermal conductivity of a granular material to the conductivities and fractional volumes of the component phases. A simple example of such a heterogeneous system is the mixture of two materials arranged in parallel slabs. If ϕ_1 and ϕ_2 are the volume fractions of the constituent slabs with thermal conductivities λ_1 and λ_2, then the thermal conductivity of the mixture for a flow of heat parallel to the plane of the slab is given by

$$\lambda_{mix} = \phi_1 \lambda_1 + \phi_2 \lambda_2 \qquad 10.6$$

For a flow of heat perpendicular to the plane of the slabs, the conductivity is given by

$$\lambda_{mix} = \frac{\lambda_1 \lambda_2}{\phi_1 \lambda_1 + \phi_2 \lambda_2} \qquad 10.7$$

In more realistic situations in actual materials, the heat flow through a uniform medium may be disturbed by a region of different conductivity. Maxwell (1904) solved the problem of randomly sized spheres of one medium randomly distributed in another medium. The conductivity of the mixture (λ_{mix}) is then given by

$$\frac{\lambda_{mix}}{\lambda_{cont}} = \frac{1 + 2x - 2\phi(x-1)}{1 + 2x + \phi(x-1)} \qquad 10.8$$

where x is the ratio of the conductivities of the continuous and dispersed phases and ϕ the fractional volume of the dispersed phase.

This model was extended to determine the conductivity of a system containing three or more different phases, such as a mixture in which spherical particles of materials with conductivities λ_1 and λ_2 are embedded in a material of conductivity λ_0. The thermal conductivity of such a three-phase mixture was obtained by Brailsford and Major (1964) assuming the phase with conductivity λ_0 to be continuous. The conductivity is given by

$$\lambda_{mix} = \frac{\left\{\lambda_0 \phi_0 + \lambda_1 \phi_1 \frac{3\lambda_0}{(2\lambda_0 + \lambda_1)} + \lambda_2 \phi_2 \frac{3\lambda_0}{(2\lambda_0 + \lambda_2)}\right\}}{\left\{\phi_0 + \phi_1 \frac{3\lambda_0}{(2\lambda_0 + \lambda_1)} + \phi_2 \frac{3\lambda_0}{(2\lambda_0 + \lambda_2)}\right\}} \quad (10.9)$$

Here ϕ_0, ϕ_1 and ϕ_2 refer to the volume fractions of the corresponding phases.

A random two-phase assembly can be regarded as having regions of both phases in the correct proportions, embedded in a random mixture of the same two phases, having a conductivity equal to the average conductivity of the two-phase assembly. This may be achieved by setting $\lambda_0 = \lambda_{mix}$ in the above equation and solving for λ_{mix}.

$$\lambda_{mix} = \{(3\phi_1 - 1) \lambda_1 + (3\phi_2 - 1) \lambda_2\} + (\{(3\phi_1 - 1) \lambda_1 + (3\phi_2 - 1) \lambda_2\}^2 + 8\lambda_1 \lambda_2)^{1/2} \quad (10.10)$$

where $\phi_2 = 1 - \phi_1$.

Sugawara and Yoshizawa (1961) measured the thermal conductivities of materials which conformed to the two-phase assembly. A comparison between the theoretical and experimental results shows good agreement.

These considerations are not specific to semiconducting materials and the systems in which this type of analysis has been usefully applied are sandstones, glass-balls surrounded by air, perforated rubber plates, etc.

In the discussions so far, the individual regions of each phase are assumed sufficiently large to be characterized by their individual thermal conductivities. However, they are small in comparison with the total volume so that small-scale behaviour of individual regions may be ignored.

10.4.2. *Random inhomogeneities in semiconductors*

A number of research papers have been published in which the effect of random inhomogeneities on the electrical properties of semiconductors have been investigated. Two of the methods are of particular importance and will be discussed here.

In semiconductors, inhomogeneities may take several forms. The impurity concentration may vary spatially. The transport coefficients may vary from point to point in the material but the basic semiconductor parameters, such as energy band gap, carrier effective-mass, etc. may remain invariant. On the other hand, there may be variations in the basic semiconductor parameters, such as in alloys where the alloy composition varies.

Herring (1960) discussed a statistical method using the small-fluctuation theory. In this method, the local transport tensors and vectors are considered as the sum of an infinite number of Fourier components. By solving the transport equations for these components, the properties of the material may be expressed in terms of the mean and variance of the transport coefficients. This theory, however, does not hold good for large fluctuations. Airapetiants (1957) employed a macroscopic approach and considered the inhomogeneous material as a mixture of a number of components, each having different transport coefficients. In such a mixture each component will consist of a large number of irregular volumes embedded in the rest of the mixture. In the isotropic case, each volume may be taken as a sphere and the surrounding material as an infinite, homogeneous and isotropic medium. By an averaging process the transport properties of the inhomogeneous material may be expressed in terms of those of the components. This method is successful even for large variations in the transport coefficients of the components but becomes very cumbersome when considering a large number of components.

A common form of inhomogeneity in which isolated inclusions are present in an otherwise homogeneous medium is not adequately described by either of these methods. Such inclusions may arise during and after material preparation by temporal variations in crystal growth conditions or by precipitation.

A general case of k different types of inclusions has been discussed by Ryden (1974). The ith component occupying a volume fraction ϕ_i of the material possesses transport coefficients σ_i, λ_i, α_i and R_i for electrical and thermal conductivities, and Seebeck and Hall coefficients. The inclusions are embedded in a medium whose respective coefficients are described by σ_c, λ_c, α_c and R_c. The inclusions are assumed to be spherical. A solution is then sought where the transport coefficients remain unchanged if the inhomogeneous material is replaced by a homogeneous medium with the transport coefficients σ_p, λ_p, α_p and R_p. The effective thermal conductivity (λ_p) of the inhomogeneous material is then given by (Ryden, 1974)

$$\lambda_p = \frac{\lambda_c \left\{ 1 + 2 \Sigma \phi_i \left(\frac{\lambda_i - \lambda_c}{\lambda_i + 2\lambda_c} \right) \right\}}{\left\{ 1 - \Sigma \phi_i \left(\frac{\lambda_i - \lambda_c}{\lambda_i + 2\lambda_c} \right) \right\}} \qquad (10.11)$$

The derivation of this equation is based on an assumption that the Seebeck coefficient has the same value for the inclusion and the continuous phase. If this condition is not satisfied, the presence of local currents is likely to change the thermal conductivity.

The consideration of this effect will change the thermal conductivity of the inclusion by an amount $\Delta \lambda_i$ given by

$$\lambda'_i = \lambda_i + \Delta \lambda_i \qquad (10.12)$$

Here λ_i is the thermal conductivity of the inclusion in the absence of the local current. Further

$$\Delta \lambda_i = \frac{2\sigma_i \sigma_c}{\sigma_i + 2\sigma_c} (\alpha_i - \alpha_c)^2 T \qquad (10.13)$$

is the additional component due to the Peltier heat transfer (Ryden, 1974).

There will be circulating currents crossing the boundaries of the inhomogeneous regions embedded in the continuous phase which will alter the value of λ_p. A similar argument gives an additional component

$$\Delta\lambda_p = \frac{2\sigma_p \sigma_c}{\sigma_p + 2\sigma_c} (\alpha_p - \alpha_c)^2 T \qquad (10.14)$$

α_p being the Seebeck coefficient of the inhomogeneous material. The thermal conductivity of the inhomogenous material is then given by

$$\lambda_p = \lambda_c \frac{\left\{1 + 2\Sigma\,\phi_i \left[\dfrac{\lambda_i' - \lambda_c}{\lambda_i' + 2\lambda_c}\right]\right\}}{\left\{1 - \Sigma\phi_i \left[\dfrac{\lambda_i' - \lambda_c}{\lambda_i' + 2\lambda_c}\right]\right\}} - \Delta\lambda_p \qquad (10.15)$$

where

$$\Delta\lambda_p = 2\sigma_c\, T \left[\Sigma\phi_i\left[\frac{\alpha^I - \alpha_c}{1 - \Sigma\phi_i\,(\lambda_i' - \lambda_c)/(\lambda_i' + 2\lambda_c)}\right]\left[\frac{3\sigma_i}{\sigma_i + 2\sigma_c}\right]\left[\frac{3\lambda_c}{\lambda_i' + 2\lambda_c}\right]\right]$$

$$\times \left\{\left[2 + \Sigma\phi_i\left(\frac{\sigma_i - \sigma_c}{\sigma_i + 2\sigma_c}\right)\right]\left[1 + 2\Sigma\,\phi_i\left(\frac{\sigma_i - \sigma_c}{\sigma_i + 2\sigma_c}\right)\right]\right\}^{-1} \qquad (10.16)$$

10.4.3 *Polycrystalline and hot-pressed anisotropic materials*

In materials such as $Bi_2\,Te_3$, $Bi_2\,Se_3$ and $Sb_2\,Te_3$, the electrical and thermal properties of single crystals are highly anisotropic with the electrical and thermal conductivities being higher along the two equivalent directions parallel to the cleavage planes than along the c-axis. In many situations one has to deal with the thermal conductivity of polycrystalline or hot-pressed materials. Due to the high degree of anisotropy, the thermal conductivities of these materials are not directly comparable to those of the corresponding single crystals. Ryden (1971) has obtained relationships between the thermal conductivities of polycrystalline and hot-pressed materials and the corresponding single crystal parameters. This is needed if a meaningful comparison between various published data has to be made.

The two methods discussed earlier to calculate the thermal conductivity of inhomogeneous materials can be applied to obtain thermal and electrical conductivities of polycrystalline and hot-pressed anisotropic materials.

In the macroscopic method (Airapetiants and Bresler, 1958) the procedure of obtaining the conductivity of a mixture in terms of the conductivities of the components is used to obtain transport coefficients of a powder compact in terms of the corresponding parameters of the single crystal. Ryden has obtained expressions for thermal conductivity of polycrystals and powder compacts using the macroscopic approach as well as that based upon the small fluctuation theory (Herring, 1960).

The thermal conductivity of a powder sample can be expressed in terms of thermal conductivities of the corresponding single crystal along the cleavage planes ($\lambda_\parallel$) and along the c-axis ($\lambda_\perp$) perpendicular to the cleavage plane. The thermal conductivity is given by

$$\lambda = \langle\lambda\rangle\left\{1 - \frac{2}{3}\left(\frac{\lambda_\| - \lambda_\perp}{2\lambda_\| + \lambda_\perp}\right)^2\right\} \quad (10.17)$$

where
$$\langle\lambda\rangle = \frac{2\lambda_\| + \lambda_\perp}{3} \quad (10.18)$$

This derivation assumes the isotropy of the Seebeck coefficients to avoid the effect of local currents. Taking the anisotropy of Seebeck coefficient into consideration the thermal conductivity of a powdered material is given by

$$\lambda = \langle\lambda\rangle\left\{1 - \frac{2}{3}\left(\frac{\lambda_\| - \lambda_\perp}{2\lambda_\| + \lambda_\perp}\right)^2 + \frac{4}{3}\frac{(\alpha_\perp - \alpha_{11})^2\langle T\rangle}{(\rho_\perp + 2\rho_\|)(\lambda_\perp + 2\lambda_\|)}\right\} \quad (10.19)$$

ρ refers to electrical resistivity.

For a polycrystalline material, the thermal conductivity is given by

$$\lambda = \langle\lambda\rangle\left\{1 - \left(\frac{\lambda_\| - \lambda_\perp}{\lambda_\| + \lambda_\perp}\right)^2 + \frac{1}{2}\frac{(\alpha_\perp - \alpha_{11})^2\langle T\rangle}{(\rho_\perp + \rho_\|)(\lambda_\perp + \lambda_\|)}\right\} \quad (10.20)$$

where $\langle\lambda\rangle = \dfrac{\lambda_\| + \lambda_\perp}{2}$ (10.21)

Ryden compared the predictions of the theory described earlier to those obtained on the basis of the macroscopic method. For large anisotropies, the macroscopic method gave more accurate results. However, for small anisotropies, the small-fluctuation theory gave good results. Fig. 10.6 gives

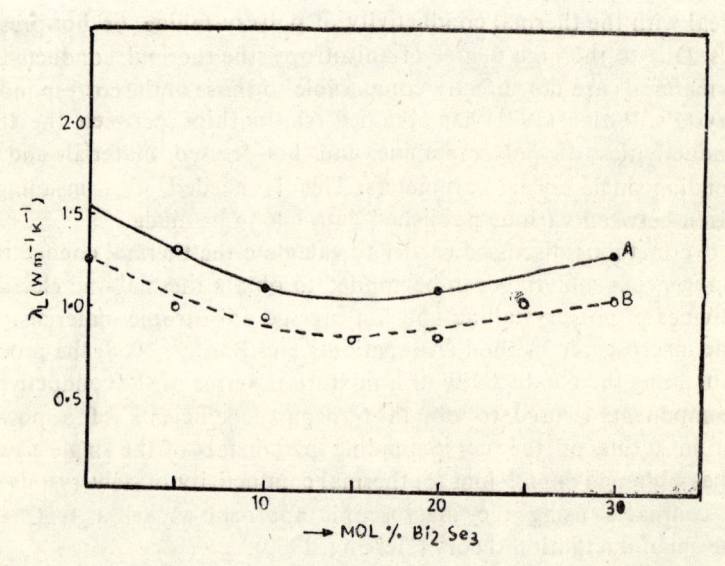

Fig. 10.6 The room-temperature thermal conductivity of Bi_2Te_3–Bi_2Se_3 as a function of composition. Curve A fits single-crystal data ($\lambda_\|$); curve B fits the theory with anisotropy ratio $\lambda_\|/\lambda_\perp = 2$. xx Goldsmid (1961), o, AERE, MCP results (after Ryden, 1971).

the room-temperature thermal conductivity of Bi_2Te_3–Bi_2Se_3 alloys as a function of alloy composition. The theoretical results (curve B) obtained using the single-crystal data are in good agreement with the experimental data on pressed-powder compacts. Ryden's analysis shows that the thermal conductivity of polycrystalline anisotropic material or pressed-powder compacts of such materials can be expressed in terms of the thermal conductivity of the corresponding single crystal.

References

Airapetiants, S.V. (1957), *Sov. Phys.-Tech. Phys.* 2, 429.
Airapetiants, S.V. and Bresler, M.S. (1958), *Sov. Phys.-Tech Phys.* 3, 1778.
Bhandari, C.M. and Verma, G.S. (1966), *Phys. Rev.152*, 731.
Brailsford, A.D. and Major, K.G. (1964). Brit. J. Appl. Phys. 15. 313.
Callaway, J. and Boyd, R. (1964), *Phys. Rev. 134*. A1655.
Douglass, R.L. (1963), *Phys.Rev.129*, 1132.
Douthett, D. and Friedberg, S.A. (1961), *Phys. Rev. 121*, 1662.
Friedberg, S.A. and Harris, E.D. (1963), *Low-Temperture Physics LT 8*, Butterworth Scientific Publ. LTD., London, p. 302.
Friedman, L. (1964), *Phys. Rev. A133*, 1668.
Friedman, L. (1965), *Third Organic Crystal Symposium*, Chicago. Illinois.
Garrett, C.G.B. (1959), *Semiconductors* (ed G.B.Hannay), Reinhold Pub. Co., New York, Chapter 15.
Goldsmid, H.J. (1961), *J. Appl. Phys., Suppl. 32*, 2198.
Gutman, F. and Lyons, L.E. (1967), *Organic Semiconductors*, John Wiley and Sons, Inc., New York.
Herring, C. (1960). *J. Appl. Phys. 31*, 169.
Kaganov, M.I. and Tsukernik, V.M. (1959), *Sov.Phys. JEPT 9*, 151.
Kittel, C. (1958), *Phys Rev. 110*, 836 (1958).
Kittel, C. (1964), *Quantum Theory of Solids*, Wiley, New York.
Kittel, C. and Abraham, E. (1953), *Rev. Mod. Phys. 25*, 233.
Klemens, P.G. (1977), *Proc. of 7th Symposium on Thermophysical Properties* (ed. A. Cezairliyan), New York, ASME.
Luthi, B. (1962), *J. Phys. Chem. Solids 23*, 35.
Maxwell, J.C. (1904), *A Treatise on Electricity and Magnetism*, Clarendon Press, Oxford.
McCollum, D.C., Wild, R.L. and Callaway, J. (1964), *Phys. Rev. 130*, 4420.
Mogilevskii, E.M. and Surin, V.G. (1972), *Sovt Phys.- Sol. State 13*, 2071.
Pratt, A.W. (1969), *Thermal Conductivity* (ed. R.P. Tye), Vol 1, Academic Press, London.
Rezanov, A.V. and Cherepanov, V.J. (1953), *Proc. Acad. Sci. USSR 93*, 641.
Ross, R.G., Andersson, P. and Backstrom, G. (1979 I), *Mol. Phys.* 38, 377.
Ross, R.G., Andersson, P. and Backstrom, G. (1979 II), *Mol. Phys.* 38, 527.
Ryden, D.J. (1971), *J. Phys. C: Sol. St. Physics 4*, **1193**
Ryden, D.J. (1974), *J. Phys. C: Sol. St. Physics 7*, 2655.

Sanders. D.J. and Walton, D. (1977), *Phys. Rev. B 15*, 1489.
Sato, H. (1955), *Progr. Theor. Phys.* (Kyoto) 13, 119.
Siebrand, W. (1962), *Organic Crystal Symp.* (NRC Ottawa); also *Doctorate Thesis*, University of Amsterdam (1963).
Sinha, K.P. and Upadhyaya, U.N. (1962), *Phys. Rev. 127*, 432.
Slack, G.A. (1961), *Phys. Rev. 122*, 1451.
Slack, G.A. (1962), *Phys. Rev. 126*, 427,
Slack, G.A. (1964), *Phys. Rev. 134*, A1268.
Slack, G.A. and Galginaitis, S. (1964), *Phys. Rev. 133*, A253.
Slack, G.A. (1979), *Solid St. Physics* (ed. Ehrenreich *et al.*), Vol. 34.
Sugawara, A. and Yoshizawa, Y. (1961), *Australian J. Phys. 14*, 469.
Szent-Gyorgyi, A. (1946), *Nature 157*, 875.
Uberreiter, K. and Orthmann, H.J. (1950), *Z Naturforsch 5a*, 101.
Van Kranendonk, J. and Van Vleck, J.H. (1958), *Rev. Mod. Phys. 30*, 1.
Walton, D., Rives, J.E. and Khalid, Q. (1973), *Phys. Rev. B 8*, 1210.

Chapter XI

Applications of Thermal Conductivity Studies and Other Topics

11.1 Introduction

The study of thermal conduction in solids requires a knowledge of a number of properties of the material under investigation. On the other hand, thermal conductivity investigations can yield useful information on several other properties of the material. The lattice thermal conductivity which forms the major part of the thermal conductivity of semiconductors provides information about various defects in the material's lattice structure, such as charged and neutral impurities, vacancies, dislocations and elastic strain. The electronic contribution to thermal conductivity can be analysed to provide information about the carrier scattering mechanisms. The nature of the energy bands also has a strong influence on the electronic thermal conductivity which, in turn, yields information regarding the extent to which the energy bands deviate from the usually assumed parabolic nature.

Other contributions to thermal conductivity, such as those due to photons and magnons are important in specific cases. The photon thermal conductivity can provide information about the absorption coefficient of the material and its dependence on temperature. The magnon contribution is significant only in magnetic materials at very low temperatures. Thermal conductivity investigations in these materials do give information about magnon scattering mechanisms although this is limited to a very narrow range of temperatures. Table 11.1 gives a schematic presentation of various types of informations that can be obtained by investigating individual contributions to the total thermal conductivity.

11.2 Practical importance of thermal conductivity studies in semiconductors

In addition to the useful information that can be obtained about a variety of properties of a solid (or a liquid) from thermal conductivity studies a knowledge of the total thermal conductivity, and the means by which it can be increased or decreased in a selective manner, is evidently of considerable practical importance.

Table 11.1
(After Ravich et al. 1970)

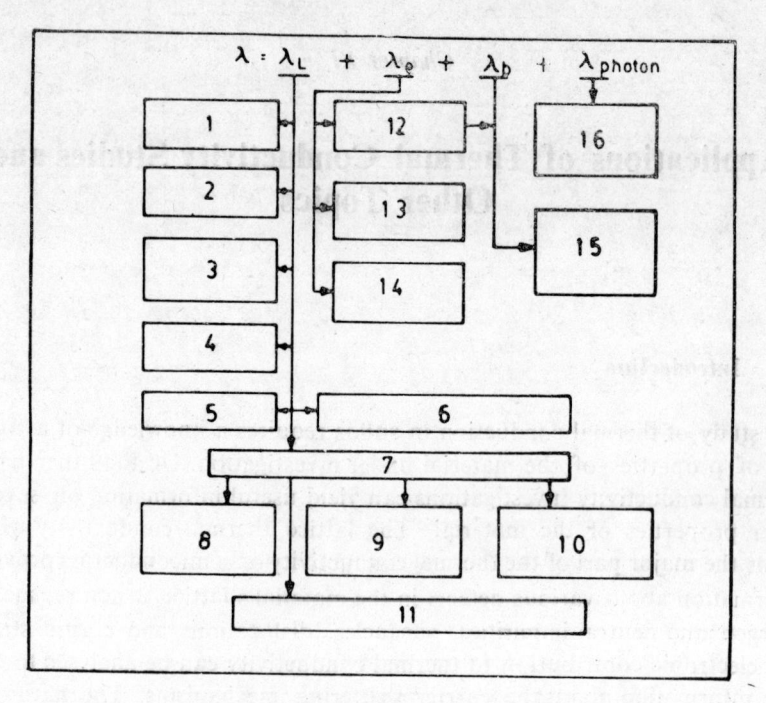

$\lambda = \lambda_L + \lambda_e + \lambda_b + \lambda_{photon}$

[Diagram with numbered boxes 1–16 showing relationships between factors affecting thermal conductivity components]

1. Kinetics of formation and precipitation of solid solutions
2. Elastic stresses in crystal lattice
3. Role of different phonon modes (optical and acoustical) in heat transport processes
4. Nature of interaction of phonons with free carriers
5. Phase transition points
6. Presence of a compound in solid solution series
7. Defects in crystal lattices
8. Location of defects in lattice (at sites or interstices)
9. Nature of defects (vacancies, impurities, dislocations, isotopes, complexes, etc.)
10. Kinetics of formation and dissociation of complexes
11. Ordered and disordered distribution of defects in lattice: kinetics of defect ordering processes
12. Structure of valence or conduction band
13. Carrier scattering mechanisms
14. Nature of interaction of electrons with phonons and other electrons
15. Absorption coefficient, and its temperature-dependence
16. Absolute value of forbidden bandwidth and its temperature-dependence

11.2.1 Thermoelectric devices

Thermoelectric generators and refrigerators employ thermoelements fabricated from semiconductor materials, and the search for potentially good thermoelectric materials has resulted in the development of a variety of semiconductors for specific applications. Thermoelectric semiconductors are heavily doped with the total thermal conductivity λ given by the sum of the lattice (λ_L) and electronic (λ_e) contributions. In a metal, the electronic thermal conductivity and electrical conductivity are related by the Wiedemann-Franz-Lorenz law while a more complicated relationship exists for semiconductors. Once the Seebeck coefficient has been optimized by suitable doping, the maximum thermoelectric figure-of-merit can be expressed in terms of the ratio λ_e/λ_L as shown in Fig. 11.1 (Rosi, 1968).

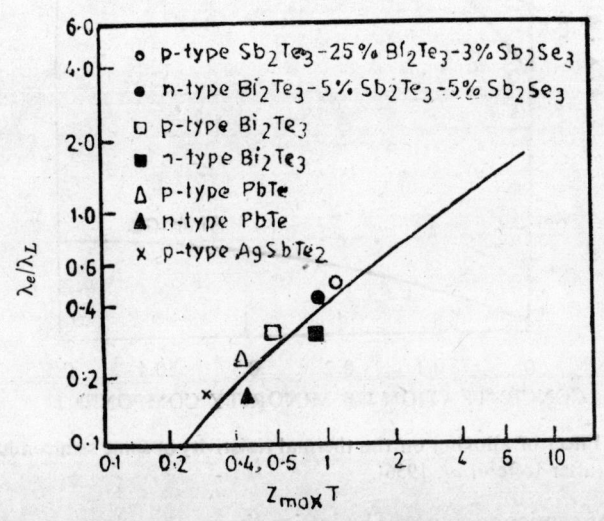

Fig. 11.1 Maximum value of dimensionless figure-of-merit ($Z_{max}T$) as a function of the ratio λ_e/λ_L for several thermoelectric materials (after Rosi, 1968).

Conventionally the figure-of-merit is related to the fundamental physical properties of the semiconductor through a so-called material parameter β' where

$$\beta' = \frac{\mu}{\lambda_L}\left(\frac{m^*}{m_0}\right)^{3/2} \qquad (11.1)$$

Here μ is the carrier mobility, m^* the density-of-states effective-mass and m_0 the free electron mass. Evidently, it is desirable for the quantity $\mu(m^*/m_0)^{3/2}$ to be as large as possible and λ_L to be small.

Substantial improvements in the thermoelectric figure-of-merit accompany the use of solid solutions (Ioffe et al., 1956) In these materials the long-range periodicity of the crystal lattice is maintained, but the short-range disturbance which results from atoms of different masses in adjacent lattice sites is effective in scattering phonons. This results in the solid solution having a thermal conductivity which is, in some cases, an order of magnitude smaller than that of its constituents as, for example, in silicon-germanium alloys. Figure 11.2 shows the effect of alloying on the thermal conductivity

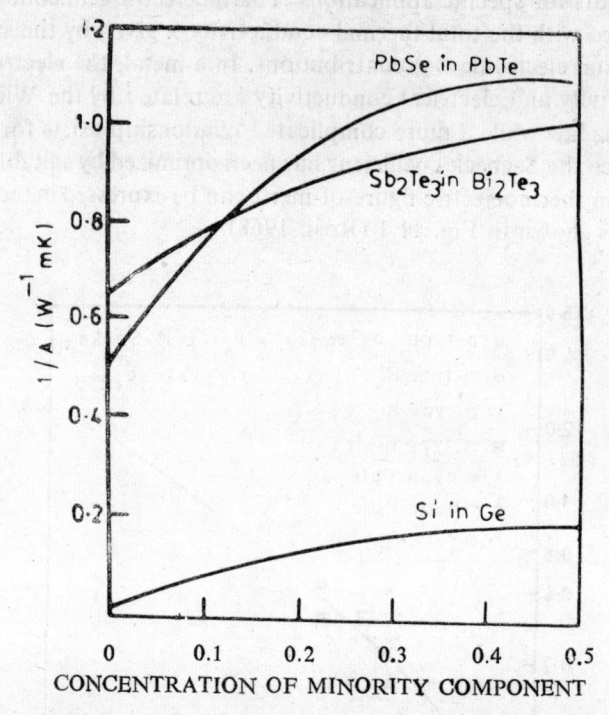

Fig. 11.2 Effect of alloying on the thermal resistivity of some semiconductors (after Ioffe *et al*, 1956).

of a number of semiconductors. The lattice thermal conductivity of an alloy is further reduced in fine-grain material, an effect attributed to phonon scattering at grain-boundaries (Goldsmid and Penn, 1968, and Parrott, 1969). This topic has been discussed in detail in Chap. VI.

Thermoelectric materials should possess a low value of lattice thermal conductivity and a relatively large energy gap to inhibit intrinsic conduction. Normally materials with large energy gaps have high values of λ_L. A plot of the energy gap E_g versus λ_L for a number of semiconductors is displayed in Fig. 11.3. The values of λ_L for Si, Ge, Si-Ge alloy crystals and sintered Si-Ge alloys demonstrate the gradual reduction in λ_L which results from an attempt to obtain more efficient thermoelectric materials.

11.2.2 Other applications

High-thermal-conductivity requirements
Study of thermal conductivity has importance in several other device applications. The requirement of devices, such as power-transistors, solar cells

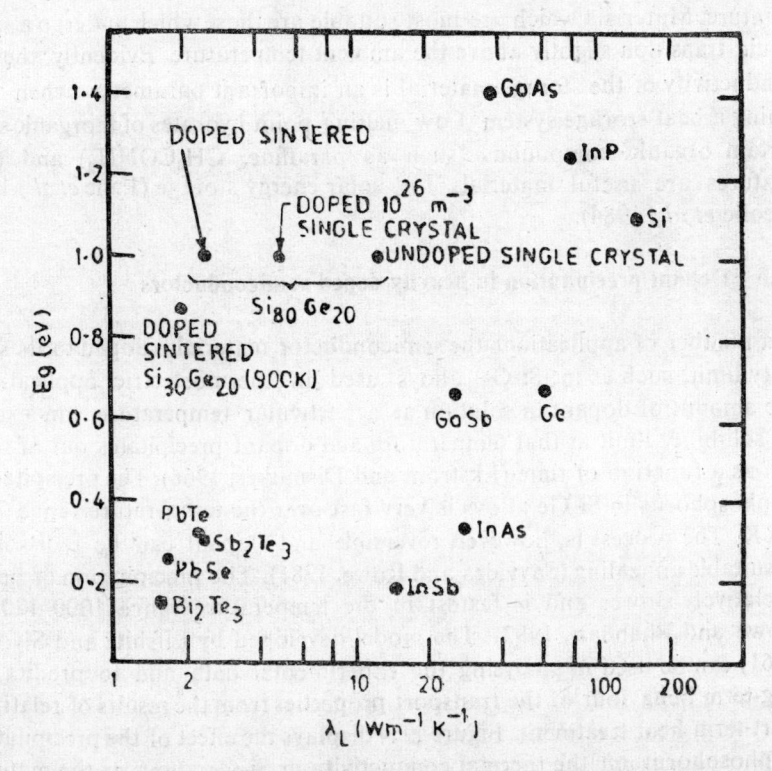

Fig. 11.3 Lattice thermal conductivity and energy gap for some semiconductors. The effect of alloying and sintering in Si-Ge alloys shows substantial reduction in λ_L (after Bhandari and Rowe, 1978 b).

and IMPATT diodes, is rather different. In these applications, the semiconductor should have a high thermal conductivity to assist the transference of heat to the sink.

Geophysical applications

A study of the thermal conductivity of the layers of the earth's mantle is of importance as most of heat transfer in this part of the earth's interior ultimately occurs by conduction although convective processes may be significant in the deeper layers. To understand the dynamic processes of the interior of the earth a study of the transport properties (including thermal conductivity) of rocks and minerals at high pressure may be of interest (Roufosse and Jeanloz, 1983). Some aspects of the thermal conductivity at high pressure are described in Sec. 11.5.

Application in heat storage systems

Thermal conductivity studies provide important information in the context of energy storage. Utilization of latent heat is one of the methods used in the storage of low-grade thermal energy. The solid-liquid phase transition facilitates the achievement of high storage densities at nearly constant temperature. Materials which are most suitable are those which undergo a solid-liquid transition slightly above the ambient temperature. Evidently, thermal conductivity of the storage material is an important parameter when designing a heat storage system. Low melting point hydrates of inorganic salts, certain organic compounds (such as paraffine, CH_3CONH_2) and their mixtures are useful materials for solar energy storage (Lane et al., 1975, Nicolic et al., 1984).

11.3 Dopant precipitation in heavily doped semiconductors

In a number of applications the semiconductor material is doped to its solubility limit, such as in Si-Ge alloys used in thermoelectric applications. The amount of dopant in solution at a particular temperature can exceed the solubility limit at that temperature and dopant precipitates out of solution as a function of time (Ekstrom and Dismukes, 1966). The precipitation of phosphorus in Si-Ge alloys is very fast over the temperature range 700–800 K. The process is, however, reversible and dopant can be redissolved by suitable annealing (Savvides and Rowe, 1981). The precipitation of boron is relatively slower and is fastest in the temperature range 1000–1200 K (Rowe and Bhandari, 1983). The model developed by Lifshitz and Slyozov (1961) can be used in analysing the experimental data and to predict the long-term behaviour of the transport properties from the results of relatively short-term heat treatment. Figure 11.4 displays the effect of the precipitation of phosphorus on the thermal conductivity of Si-Ge alloys as the material is isothermally heat-treated at 773 K.

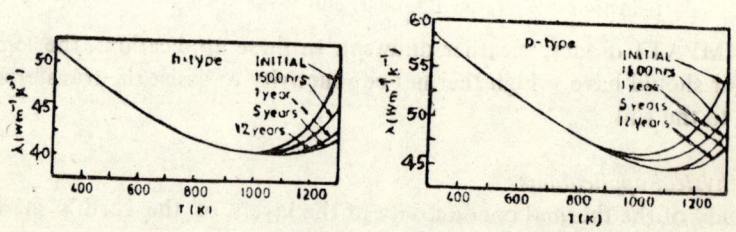

Fig 11.4 Long-term thermal conductivity behaviour of Si-Ge alloys (after Raag, 1978).

11.4 Nonmetallic crystals with high thermal conductivity

It was earlier pointed out (see Chap. I) that diamond and some other crystals with a similar structure possess thermal conductivities comparable to that of metals. Slack (1973) reviewed the experimental data on the thermal conductivity of a number of nonmetallic crystals which were classified as high-thermal-conductivity materials.

Equation 6.31 can be taken as the starting point in an analysis of the effects of various parameters on the magnitude of λ. At room temperature, the factor $\bar{M}\delta\theta_D^3\gamma^{-2}$ will determine the magnitude of λ with low γ values

APPLICATIONS OF THERMAL CONDUCTIVITY STUDIES 215

corresponding to higher values of λ. For adamantine crystals like Ge, Si and diamond, γ has the smallest values. However, because of a lack of accurate data on γ for all the crystals this term is usually omitted and the quantity $\overline{M}\delta\theta_D^3$ is taken as a convenient scaling parameter for the adamantine compounds at 300K. Slack (1973) has tabulated values of $\overline{M}$, δ, θ_D and the scaling paremeter $\overline{M}\delta\theta_D^3$ for a number of adamantine crystals. Fig. 11.5

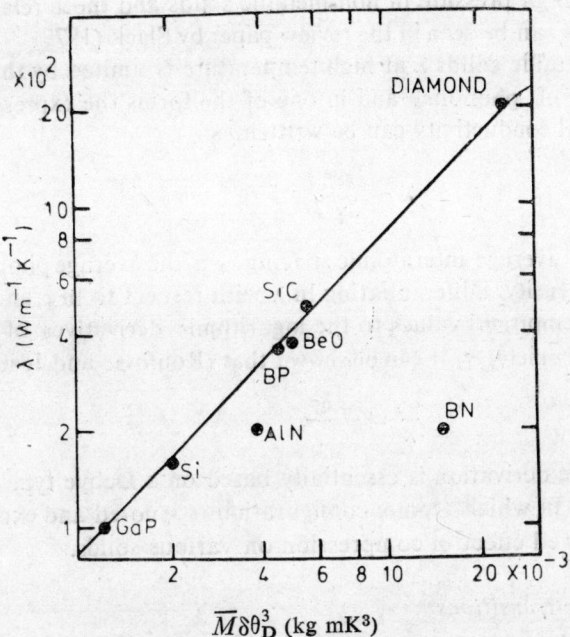

Fig 11.5 The thermal conductivity of various adamantine compounds against the scaling parameter (after Slack, 1973).

shows a plot of λ versus the scaling parameter. The agreement is good except for AlN and BN; Slack attributed the lower values of λ in these cases to the presence of impurities.

There are about twelve adamantine crystals with $\lambda > 100$ Wm^{-1} K^{-1} at room temperature. λ for high purity diamonds (both synthetic and natural) may be as high as 2000 Wm^{-1} K^{-1}. Some semiconductors such as Si, GaP belong to this category of high-thermal-conductivity materials. Slack has outlined the criteria for the nonmetallic crystals to possess a high thermal conductivity. These are: low average atomic mass, strong interatomic bonding, a simple crystal structure, and low anharmonicity.

11.5 Thermal conductivity at high pressure

The effect of pressure on lattice thermal conductivity is usually interpreted by considering its effect on the Debye temperature and phonon velocity. Both these quantities increase with an increase in pressure and, therefore, the thermal conductivity is expected to increase with pressure. Moreover, the lattice constant decreases with compression and this results in an increase in the size of the Brillouin zone. However, this effect can usually be ignored as the change in the lattice constant is relatively small (Hughes and Savin, 1967) and in many cases the observed changes in the thermal conductivity of non-metallic solids can be interpreted without taking lattice constant changes into account. References to various papers related to the measurement of λ at high pressure in non-metallic solids and those related to theoretical studies can be seen in the review paper by Slack (1979).

In non-metallic solids λ at high temperature is limited by the anharmonic coupling of phonons, and in one of the forms the expression for the lattice thermal conductivity can be written as

$$\lambda \sim \frac{r\bar{v}^3}{3\gamma^2 T} \rho \qquad (11.2)$$

where r is the average interatomic spacing, $\bar{v}$ is the average phonon velocity and ρ is the density. Differentiating $\ln \lambda$ with respect to $\ln \rho$ and assigning approximate empirical values to the logarithmic derivatives of $\bar{v}$ and the Gruneisen parameter γ, it can be shown that (Roufosse and Jeanloz, 1983)

$$\frac{\delta \lambda}{\lambda} \simeq 7 \frac{\delta \rho}{\rho} \qquad (11.3)$$

This simple derivation is essentially based on a Debye type of "continuum model" in which atomic configuration is ignored and explains fairly well the observed effect of compression on various solids.

Effect of phase-transitions

The application of pressure brings about phase-transitions in some cases. In some transitions, such as that across the quartz-coesite and olivine-spinel transitions, the continuum model reproduces reasonably well the effect of compression on λ.

However, in certain phase-transitions such as coesite-stishovite, B_1—B_2 (in KCl and RbBr) and pyroxene-garnet transitions, the observed effects of compression deviate in a significant way from the theoretical predictions based on the simple model and show a lower conductivity for the high-density phase contrary to usual expectations (Yukutake and Shimada, 1974, Averkin *et al*, 1977, and Osako and Kobayashi, 1979).

The structural contribution to thermal conductivity must, therefore, be calculated to account for this type of behaviour. This has to be explicitly calculated for each crystal structure as it does not follow a simple pattern. Roufosse and Jeanloz (1983) have described the details of these calculations and shown that the change in λ in a B_1—B_2 transition is due mainly to a change in the number of atoms in the unit cell which results in a decrease in the thermal conductivity of the high-pressure phase.

Little work has been reported on the high-pressure thermal conductivity of semiconductors. Amirkhanova *et al.* (1976) measured the high-pressure thermal conductivity of Cd_xHg_{1-x} Te, where $x = 0.28$ and these results can be interpreted on the basis of the continuum model. A study of the effect of hydrostatic pressure up to 3.5 Kbar on the thermal conductivity of *n*-type GaSb was made by Amirkhanova *et al* (1983). The thermal conductivity of monocrystalline GaSb as a function of temperature under the influence of hydrostatic pressure was found to increase linearly from 10 to 22 per cent.

11.6 Radiation damage in solids—effect on thermal conductivity

11.6.1 *Introduction*

The technological revolution during the post-war years has placed increasingly severe demands upon material research. A number of advanced-technology applications require new materials with properties which ensure that they maintain their performance when subjected to abnormal operating environments. Nuclear stability is one such requirement with materials used in nuclear reactors being developed to withstand better high temperatures and high radiation fluxes for prolonged periods of time. Evidently, there is need for detailed studies of the effects of radiation on the relevant material properties.

Radiation damage includes any change in material properties which results from the interaction of radiation with matter. However, in the context of thermal conduction studies, the discussion will be confined to the damage produced in crystalline lattices when exposed to energetic radiations. The term "damage" does not necessarily mean that the properties of the material under investigation will be adversely affected, and in a number of instances irradiation results in the material acquiring some desirable properties. The change in physical properties, desirable or otherwise, often arises from the ability of radiation to displace atoms from their equilibrium positions and destroy the order of the crystalline lattice.

Initially the study of radiation damage in solids arose from the engineering requirements of nuclear reactor technology. However, radiation produced defects have now become an important tool in the study of the solid state (Billington and Crawford, 1961).

11.6.2 *Radiation damage in quartz—effect on thermal conductivity*

Berman (1951) studied the effect of neutron bombardment on the low-temperature thermal conductivity of quartz and his results are shown in Fig. 11.6. With increased radiation, the thermal conductivity decreases progressively over the entire temperature range of 2–100 K and the conductivity maximum becomes less sharp. For the highest exposures, the thermal conductivity approached the value corresponding to fused silica. This is expected as the minimum of the phonon mean-free-path will correspond to a situation in which it is about the dimensions of a phonon wavelength (see Sec. 6.11).

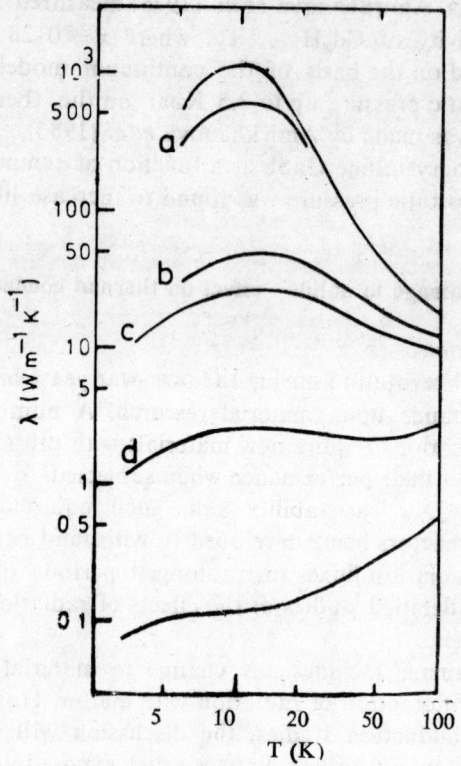

Fig 11.6 Thermal conductivity of quartz crystal, irradiated quartz and fused silica. Top curve-before irradiation; curve a-after 0.03 units exposure; curve b-after 1 unit; curve c-after 2.4 units; curve d-after 19 units (1 unit is $1.8 \times 10^{22} m^{-2}$), Lower curve-fused silica (after Berman, 1951).

A theoretical analysis of this data was provided by Klemens (1951 and 1956) who assumed the extra resistance due to radiation damage (W_{RD}) to be additive. For temperatures above the conductivity maximum, W_{RD} increases approximately linearly with temperature. This type of behaviour is expected from point-defects. On the other hand below the conductivity maximum, W_{RD} is found to increase with decreasing temperature. This was interpreted in terms of phonon scattering from disordered regions extending to around 100 atomic distances.

Some important conclusions can be drawn from these and other studies related to the absorption-edge shifts and heat capacity measurements (see for example, Billington and Crawford, 1961). Up to about 3×10^{23} n_f m^{-2} (n_f refers to neutron flux) the damage is mainly explained in terms of point-defects and small regions of disorder. At higher exposures the concentration of disordered regions increases considerably and shearing stresses are produced which distort the lattice. The point-defects and ruptured covalent bonds form a network which resembles a glassy structure. At high exposures ($> 1.2 \times 10^{24}$ n_f m^{-2}), the long-range order is totally destroyed.

Table 11.2
Effect of Radiation on Thermal Conductivity and Density

Material	Initial Value Thermal conductivity cal m⁻¹ s⁻¹ K⁻¹ ×10⁻²	Density kg m⁻³ ×10³	Exposure 1 n_f (m⁻²) ×10²³	After Exposure 1 Thermal conductivity cal m⁻¹ s⁻¹ K⁻¹ ×10⁻²	Density kg m⁻³ ×10³	Exposure 2 n_f (m⁻²) ×10²³	After Exposure 2 Thermal conductivity cal m⁻¹ s⁻¹ K⁻¹ ×10⁻²	Density kg m⁻³ ×10³
Sapphire	600±200	3.983	6	300±60	3.969	60	200±30	3.914
Al₂O₃ sintered	400±100	3.559	3	230±40	3.553	40	90±5	3.80
BeO	600±200	2.84	7	400±100	2.85	40		
Spinel	250±50	3.60	7	130±10	3.60	40	130±10	3.60
Forsterite	250±50	3.056	6	75±10	3.03			
Zirconi	120±10	3.73	5	23±1	3.48	30		3.38
Steatite	76±5	2.796	7	28±1	2.760			
Corderite	73±5		5	20±2		30	20±2	
TiO₂	165±20	4.01	6	110±10	3.99	30	65±5	3.98
Porcelain	270±50	3.41	6	120±10	3.40	40	85±5	3.39
Mica	17±1	2.845	4	12±1	2.738	20	28±3	2.444
Plate glass	25±1	2.509	3		2.530	60		2.515
Silica glass	35±1	2.204	7		2.255	40		2.23

After Billington and Crawford (1961)

11.6.3 Other materials

There has been considerable effort towards understanding the nature of the radiation-induced defects in various solids including semiconductors (Whitehouse, 1973, and Albany, 1979). However, the effcts of radiation damage on thermal conductivity is usually understood in terms of additional point-defect scattering as described for the case of quartz. The thermal conductivity of neutron-irradiated silicon was measured by Savvides and Goldsmid (1974) for thin specimens and the results analysed by including point-defect and boundary scattering along with the three-phonon processes (Bhandari and Rowe 1978, also see chap. VI).

Some other materials such as sapphire, Al_2O_3 and BeO show large decrease in thermal conductivity after exposure even though the changes in density are relatively small (Table 11.2).

References

Albany, H.J. (1979), Defects and Radiation Effects in Semiconductors, *Int. Conf.*, Nice, Sept 1978, The Inst. of Physics, London.

Amirkhanova, Kh. I., Magomedov, Ya. B., Emirov, S.N. and Gadzhieva, R.M. (1976), *Sov. Phys.-Solid St. 17*, 1956.

Amirkhanova, Kh. I., Kramynina, N.L. and Emirov, S.N. (1983), *Sov. Phys.-Solid St. 25*, 1427.

Averkin, A.A., Logachev, Y.A , Petrov, A.V. and Tsybkina, N.S , (1977), *Sov. Phys.-Solid St. 19*, 988.

Berman, R. (1951), *Proc. Roy. Soc. A208*, 90.

Bhandari, C.M. and Rowe, D.M. (1978 a), *J. Phys. C: Solid St. Phys. 11*, 1787.

Bhandari, C.M. and Rowe, D.M. (1978 b), *Int. Conf. on Thermoelectric Energy Conversion*, University of Texas at Arlington, p. 32.

Billington, D.S. and Crawford, J.H., Jr. (1961), *Radiation Damage in Solids*, Princeton University Press.

Ekstrom, S. snd Dismukes, J.P. (1966), *J. Phys. Chem. Sol. 8*, 57.

Goldsmid, H.J. and Penn, A.W. (1968), *Phys. Lett. 27A*, 523.

Hughes, D.S. and Savin, F. (1967), *Phys. Rev. 161*, 861.

Ioffe, A.P., Airapetiants, S.V., Ioffe, A.V., Kolomoets, N.V. and Stil'bans, L.S.(1956), *Dokl. Akad. nauk. SSSR 106*, 981.

Klemens, P.G. (1951), *Proc. Roy. Soc. A208*, 108.

Klemens, P.G. (1956), *Phil. Mag. 1*, 938.

Lane, G.A. et al , (1975), *Proc. of the Workshop on Solar Energy Storage for the Heating and Cooling of Buildings*, NSF-RA-N-041.

Lifshitz, I.M. and Slyozov, V.V. (1961), *J. Phys. Chem. Solids 19*, 35.

Nicolic, R., Kelick, K. and Neskovic, O. (1984), *Appl. Phys. A34*, 199.

Osako, M. and Kobayashi, W. (1979), *Phys. Earth Planet Inter. 18*, 1.

Parrott, J.E. (1969), *J. Phys. C: Solid St. Phys. 2*, 147.

Raag, V. (1978), *Proc. of Second Int. Conf. On Thermoelectric Energy Conversion*, Univ. of Texas at Arlington, p. 5.

Ravich, Yu. I., Efimova, B.A. and Smirnov, I.A. (1970), *Semiconducting Lead Chalcogenides* (ed. L.S. Stil'bans) Plenum Press, New York.

Rosi, F.D. (1968), *Solid St. Electronics 11*, 833.

Roufosse, M.C. and Jeanloz, R. (1983), *J. Geophys. Res. 88*, 7399.

Rowe, D.M. and Bhandari, C.M. (1983), *Modern Thermoelectrics*, Holt Saunders, Ltd., London.

Savvides, N. and Goldsmid, H.J. (1974), *Phys. Stat. Solidi (b) 63*, K 89.

Savvides, N. and Rowe, D.M. (1981), *J. Phys. D: Appl. Phys 14*, 723.

Slack, G.A. (1973), *J. Phys. Chem. Solids 34*, 321.

Slack, G.A. (1979), Solid State Physics (ed. Ehrenreieh *et al*), Academic Press, Vol. 34, 1.

Whitehouse, J.E. (1973) (edited), Radiation Damage and Defects in Semiconductors, *Int. Conf. at the Univ. of Reading*, July 1972, The Inst. of Phys., London.

Yukutake, H and Shimada, M. (1974), *Proc. 4th Int. Conf. on High Pressure*, p. 362.

Subject Index

Absolute axial heat flow method, 15, 16
Acoustic phonon, 79, 105–107
Acoustic scattering, 53
Amorphous materials, 175–178
Angstrom's method, 20
Anharmonic crystal, 8, 85

Bipolar thermal conductivity, 65, 188
Bismuth telluride, 167, 205
Boltzmann equation, 83
Boundary scattering, 112, 113
Boundary scattering at high temperature, 113
Brillouin zone, 39
Burgers' vector, 91

Callaway equation, 109
Comparative methods, 17, 18
Compound semiconductors, 149

Deformation potential, 53
Density-of-states, 81
Diffusivity, 7
Directly heated electrical method, 18
Disorder parameter, 90
Disorder scattering, 90, 142
Dispersion curves, 80
Distribution function for electrons, 49
Dopant precipitation, 214
Doped semiconductors, 115

Effect of defects, 82
Effect of phase transitions, 216
Effect of thermal expansion, 123
Electron Boltzmann equation, 49
Electron—phonon interaction, 52, 53, 92–96
Electron scattering mechanism, 52
Energy bands, 39, 40, 43
Ettinghausen coefficient, 72

Entropy, 56
Excitonic thermal conductivity, 122
Extrinsic conduction, 42
Extrinsic semiconductor, 42

Fermi level, 44
Fine-grained Si-Ge alloys, 142
Free electrons, 37

Gallium antimonide, 154
Gallium arsenide, 153
Generalized flows, 48
Generalized forces, 48
Geophysical applications, 215
Germanium, 132–134
Gruneisen constant, 102

Hall effect, 71
Harmonic approximation, 8, 78
Heat flow, 7, 48, 84
Heat storage systems, 213
Heterogeneous solids, 200
High temperature thermal conductivity, 126
High-thermal-conductivity materials, 214, 215
Hole, 42
Holland's analysis, 110

Imperfect crystals, 116
Improved variational principles, 122,
Impurity conduction, 42
Indium antimonide, 149
Indium arsenide, 151
Intrinsic conduction, 41, 42
Intrinsic semiconductor, 41, 42
Ionized impurity scattering, 53
Inhomogeneities, 203,
Insulators, 8, 40
Intervalley scattering, 54

Index

Klemens—Callaway theory, 109, 110

Lead chalcogenides, 160
Liquid semiconductors, 183–188
Lorenz factor, 59, 61
Lorenz number, 59

Magnetic field effect, 158
Magnetic semiconductors, 192–198
Magnon thermal conductivity, 193–197
Mass-difference scattering, 90
Mean-free-path, 8, 180, 182
Measurement of specific heat, 25
Measurement of temperature, 13
Measuring techniques, 13
Metals, 41
Minimum thermal conductivity, 117, 182
Mixed scattering, 62
Multivalley structure, 43

Nernst coefficient, 71
Nonparabolic bands, 66
Non-steady-state methods, 20
Normal process, 85, 86, 109

Onsager relations, 48
Optical phonons, 118, 123–125
Order-disorder transformation, 171
Organic semiconductors, 198-200

Parabolic bands, 61–64
Periodic boundary conditions, 38
Periodic temperature-wave method 20
Phonon drag, 75
Phonon-phonon interaction, 85
Phonon modes, 79-80
Phonon scattering, 85-90
Photoacoustic effect, 31
Photon thermal conductivity, 121
Plateau in the $\lambda-T$ curves, 179
Polycrystalline anisotropic materials, 205
Pyroelectric methods, 32

Radial heat-flow method, 16
Radiation damage, 217
Radiative heat transfer, 139

Random inhomogeneities, 203
Random mixtures, 200
Reciprocal lattice vectors, 39
Relaxation time, 51
Resonance scattering of phonons 91, 154

Scaling parameter, 6, 165
Scattering of electrons, 52
Scattering of phonons, 84
Scattering of phonons by dislocations, 91
Scattering of phonons by electrons, 92
Scattering of phonons by boundaries, 113
Semiconductors, 41–46
Semiconductor melts, 27
Semimetal, 40, 41
Silicon-germanium alloys, 137–139
Silicon, 136
Static (steady-state) methods, 15

Thermal anomaly, 181,
Thermal comparator, 26
Thermal conductivity at high pressure, 215, 216
Thermal expansion, 123
Thermodynamic concepts, 56
Thermoelectric applications 211, 212
Thermoelectric devices, 211
Thermomagnetic effects, 71
Thin films, 28
Three-phonon processes 85, 88
Transitory methods, 23
Transport coefficients, 48
Transverse and longitudinal phonons, 107
Tunneling-site model, 181

Umklapp process, 85–87
Umklapp resistance, 102

Variational principle, 54, 99
Vibrational motion, 78

Wave vector, electron, 38, 39
Wave vector, phonon, 78–80
Wiedemann-Franz-Lorenz law, 186, 211